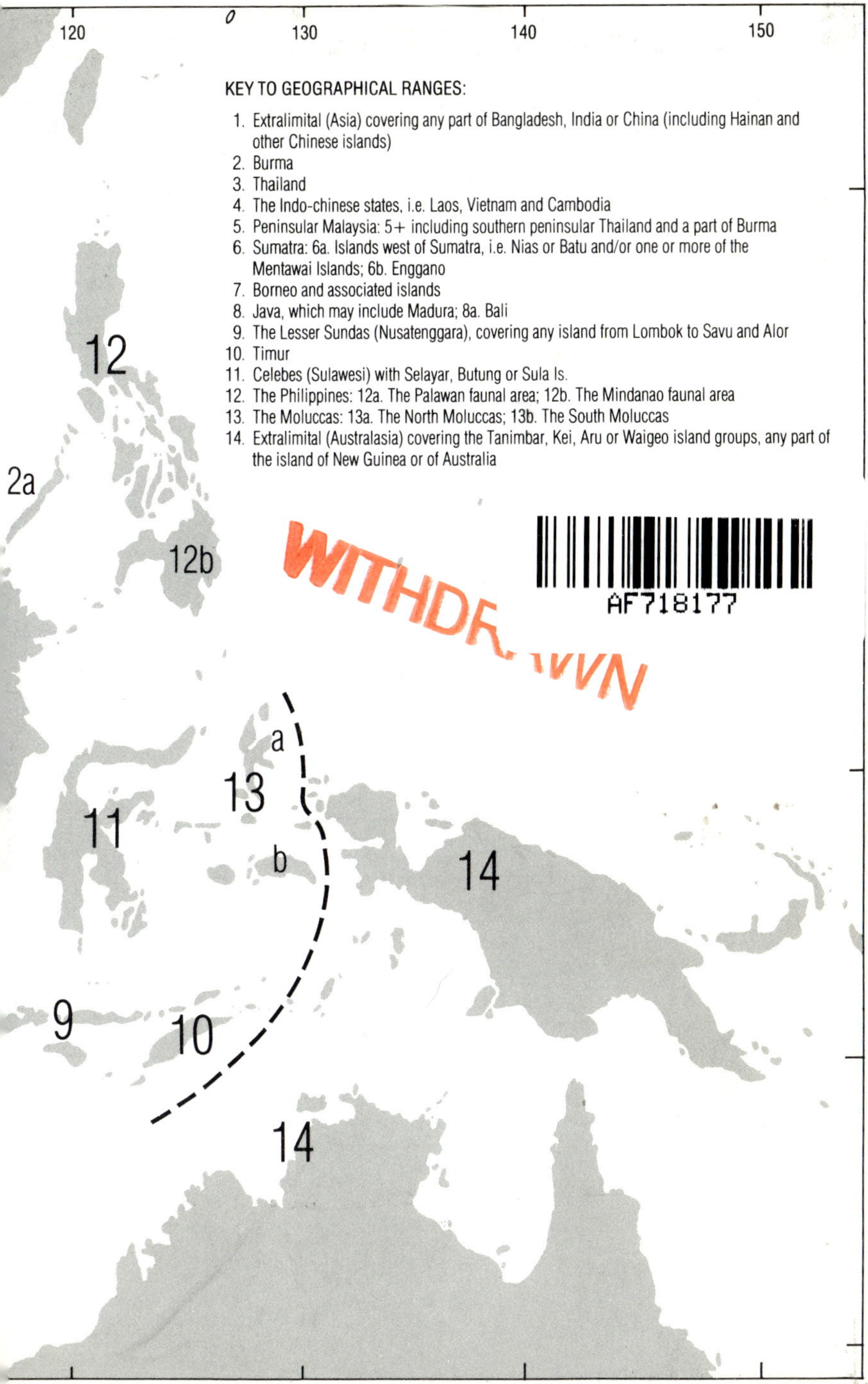

KEY TO GEOGRAPHICAL RANGES:

1. Extralimital (Asia) covering any part of Bangladesh, India or China (including Hainan and other Chinese islands)
2. Burma
3. Thailand
4. The Indo-chinese states, i.e. Laos, Vietnam and Cambodia
5. Peninsular Malaysia: 5+ including southern peninsular Thailand and a part of Burma
6. Sumatra: 6a. Islands west of Sumatra, i.e. Nias or Batu and/or one or more of the Mentawai Islands; 6b. Enggano
7. Borneo and associated islands
8. Java, which may include Madura; 8a. Bali
9. The Lesser Sundas (Nusatenggara), covering any island from Lombok to Savu and Alor
10. Timur
11. Celebes (Sulawesi) with Selayar, Butung or Sula Is.
12. The Philippines: 12a. The Palawan faunal area; 12b. The Mindanao faunal area
13. The Moluccas: 13a. The North Moluccas; 13b. The South Moluccas
14. Extralimital (Australasia) covering the Tanimbar, Kei, Aru or Waigeo island groups, any part of the island of New Guinea or of Australia

IMAGES OF ASIA

General Editor: MICHAEL SMITHIES

Riches of the Wild

Other titles in the series

At the Chinese Table
T. C. LAI

Balinese Paintings
A. A. M. DJELANTIK

Chinese Jade
JOAN HARTMAN-GOLDSMITH

A Garden of Eden: Plant Life in South-East Asia
WENDY VEEVERS-CARTER

The House in South-East Asia
JACQUES DUMARÇAY

Indonesian Batik: Processes, Patterns and Places
SYLVIA FRASER-LU

The Kris: Mystic Weapon of the Malay World
EDWARD FREY

Macau
CESAR GUILLEN-NUÑEZ

Old Bangkok
MICHAEL SMITHIES

Sailing Craft of Indonesia
ADRIAN HORRIDGE

Prince Albert's spotted Rusa, in Panay Island, the Philippines.

Riches of the Wild

Land Mammals of South-East Asia

EARL OF CRANBROOK

With coloured plates by
CDR. A. M. HUGHES, RN (RETD.)

SINGAPORE
OXFORD UNIVERSITY PRESS
OXFORD NEW YORK
1987

Oxford University Press

Oxford New York Toronto
Petaling Jaya Singapore Hong Kong Tokyo
Delhi Bombay Calcutta Madras Karachi
Nairobi Dar es Salaam Cape Town
Melbourne Auckland

and associates in
Beirut Berlin Ibadan Nicosia

OXFORD is a trademark of Oxford University Press

ISBN 0 19 582697 3

Printed in Singapore by Kyodo-Shing Loong Printing Industries Pte. Ltd.
Published by Oxford University Press Pte. Ltd.,
Unit 221, Ubi Avenue 4, Singapore 1440

Preface

ALTHOUGH it appears as an Appendix, the Checklist of land mammals of South-East Asia prepared for this book serves as an essential guide and reference to which the reader will constantly need to turn. The first chapter shows how the past geological history and present climate of this region have favoured the evolution of an exceptionally rich fauna of some 660 species. The six subsequent chapters illustrate the astonishing diversity of these mammals, group by group.

Included among uniquely South-East Asian mammals are some of the world's most primitive, notably the gymnures and treeshrews. These serve as models for extinct forms of sixty million years ago or more. In the tall evergreen rain-forests, gliding and flying mammals are abundant, including a marsupial glider, the strange, shy colugos, many flying squirrels and–most numerous of all–the bats. Bats, accounting for about one-third of all mammal species of the region, show many unusual specializations for their aerial existence. No less intriguing are the primates, which include the minute tarsiers and the huge orang-utan, confined to South-East Asia. The carnivores have diversified to occupy many niches, from the arboreal frugivore to the large ground predator, and their conservation raises several controversial issues. The ungulates are even more obviously in conflict with plantation and agricultural interests. Several large species are on the verge of extinction locally, and many unique forms are threatened. Others have proved amenable to domestication, and this may prove the route to their preservation.

The anteaters, among the strangest of mammals, are imitated by a number of rodents which depend on an insectivorous diet. Other rodents are generalized or specialized vegetarians.

Squirrels are numerous on the continent and islands of the continental shelf, but deficient or absent elsewhere. Throughout the region, but especially on islands, rats have evolved many diverse and unusual forms, some of which are convergent with squirrels in many features. The list ends with a brief mention of one of the least known and most mysterious of creatures, the Sumatran hare.

This book is the first in mammalogy devoted to South-East Asia. Based on three decades of personal involvement in the region, it is written in non-specialist language for the benefit of the general reader. I hope it will be enjoyed by all naturalists who share the author's interest in mammals and other South-East Asian wildlife. It should be especially useful to the tourist or traveller, who will find hints on good sites for mammal-watching scattered through its pages.

It is particularly gratifying to publish, at last, the coloured paintings based on a set originally commissioned by the late Tom Harrisson, former Curator of the Sarawak Museum, found among other papers passed to me by his executors after his death. The work of the artist, A. M. Hughes, is already admired through his plates for B. E. Smythies' *Birds of Burma* and *Birds of Borneo*. As far as I know, these pictures represent Commander Hughes' only venture into mammal illustration. Acknowledgement for photographs are due as follows: Plates 1 and 2, D. D. Davis; Plates 3, 17 and 18, Jane Burton; Plates 4, 7, 9, 10, 16, 19, 20, Sarawak Museum; Plate 5, K. B. Tan; Plate 6, A.N. Start; Plates 8 and 12, author; Plate 11, D. J. Chivers; Plate 13, T. C. Whitmore; Plates 14 and 15, C. R. Cox. Figure 2 is adapted from D. J. Chivers (ed.), *Malayan Forest Primates* (1980), Fig. 1.6; Figure 3 is from Alfred Russel Wallace, *The Malay Archipelago*, 1869 (American edition), p. 276, and Figure 4 is reproduced by courtesy of the American Museum of Natural History.

Saxmundham, Suffolk CRANBROOK
1986

Contents

I

Mammals and the Environment

MAMMALS are familiar: furred animals, large or small, warm-blooded, suckling their young on milk, toothed or tusked according to their kind and mostly four-limbed although variously equipped with hands, hoofs, claws, wings, or other usefully modified appendages. Domestic mammals–cattle, horses, cats, dogs, for instance–are well known. This book deals with the far greater number of wild mammals sustained by the natural environment, often as wary of man as he is ignorant of them.

South-East Asia, from Burma to the Moluccas (see end-paper map) is the home–at present count–of 660 different known and named kinds ('species') of wild land mammals (see the Checklist, Appendix), almost one-sixth of the world's total. To understand why this region is blessed with riches disproportionate to its area, we must review briefly some aspects of its history, geography, climate and natural vegetation.

The Natural Setting

In the north-west of the region, the neck of the Burma–Thai–Indo-china peninsula (from Moulmein to the Gulf of Tonkin) forms an ample connection with the heartland of Asia. On a geological time-scale, there has been no impediment to invasion and the exchange of breeding stock between wild populations. Across this border some species are shared. Many others in South-East Asia have evolved distinctive differences, yet still show clear affinity with mammals of adjacent parts of the continent.

The continental mainland tapers to the Malaysian peninsula beyond which sea separates the multitude of islands, great and small. The peninsula and islands of the continental shelf at present are separated by the shallow waters of the southern South China Sea. During the recurrent ice ages of the Pleistocene and earlier Tertiary eras, world-wide lowering of the oceans periodically exposed large parts of the shelf as dry land. In the Recent era, local sea-levels at first rose to a maximum about 6 m above the present mean, some 5,000 years ago, and have since receded. Thus, over these millennia, the area of land alternately contracted and fragmented and then expanded and reunited, creating ideal conditions for mammalian evolution. Not unexpectedly, the mammal faunas of the Sunda Shelf lands (including Peninsular Malaysia) show many similarities one with another and form a distinct subregional assemblage, here termed 'Sundaic'.

The further islands are separated from the continent by deep marine trenches: the Makassar Straits and Sulu Sea both exceed 1,000 fathoms. The sea barrier has impeded colonization from mainland sources. While major groups of mammals are therefore missing from these islands, the successful invaders of the past have given rise to a great diversity of unique and specialized forms, often confined to single islands or archipelagos. At the south-eastern extreme there is no land connection, but the mammals of Celebes (Sulawesi) and the Moluccas include an admixture of species with New Guinea/Australian affinity.

The region is one of warm climate and high rainfall. Topography creates local climatic anomalies (in the Philippines and northern Vietnam exacerbated by typhoons) but, broadly speaking, in the equatorial zone temperatures are equable and rain copious and regular. Towards higher latitudes, both north and south, an annual cycle progressively becomes dominant, dividing the year with increasing rigour into wet and dry seasons, dominated by the monsoons.

In its natural state the land is tree-covered, from the shore to mountain summits. On muddy coasts, mangroves extend down to the midtide line while, inland, good soils support the most magnificent forests of the world. In the humid equatorial belt the rain-forests are evergreen. The dense canopy absorbs about 98 per cent of the sunlight falling upon the tree-tops and casts a deep, perennial shade. In the absence of sunlight, little photosynthetic activity is possible in the shaded lower layers of the forest.

In the monsoon zone, on the other hand, many trees are deciduous. There, each year the ground is exposed for a season to the sun's radiant energy, and at the same time enriched by the fall of leaves. As the climatic transition is gradual, so too does one type of forest merge into the other. But the ecological contrast between the regimes is profound and is reflected in the mammal faunas.

In hill country, other striking changes in vegetation are associated with increasing altitude as lowland forest is replaced by upland types. Similar changes also appear among the smaller mammals, with different species adapted to restricted altitudinal ranges. Being poorly equipped to cross the intervening and (to them) inhospitable lowlands, many highland small mammal species are confined to single peaks or isolated upland blocks. On a world scale, some are extremely rare.

Conservation

During the last couple of centuries, throughout South-East Asia, man has been pushing back the original forest mantle at a progressively increasing rate. Historically, at first farm usage was transient, as it is still in the hilly interior. On good soils, if clearings are abandoned after a short period of cultivation, forest can regenerate and in time revert to near-pristine diversity. With growing human populations, however, most fertile lowlands have long been converted to permanent agriculture. Much land of lower quality has also been brought into use, often planted with perennial tree crops originating from other parts of the world (for example, rubber from South America, oil palm from West Africa). In many parts of the region, large tracts have been permanently degraded to mixed scrub, open savannah or permanent grass–vegetation types which are alien to the ancient environment.

Domestic mammals are suited to managed agricultural land. But for wild mammals that have evolved over the ages in primeval forests, such changes are disastrous. The adaptations that fit them to life among trees mainly prove impediments in other surroundings. The few that successfully exploit agricultural or plantation habitat become unwanted pests. Less flexible specialist types simply disappear. The opening up of forest land in South-East Asia has thus entailed the incidental elimination of many native wild mammals whose survival is incompatible with development.

Quantification of the loss is only possible by reference to population levels in undisturbed natural environment. Some very large numbers have been calculated: for instance, 828,000 monkeys and gibbons lost through forest clearance in Peninsular Malaysia over the years 1958–75. On islands and mountains localized endemic populations cannot be replenished by reinvasion, and habitat loss can reduce numbers beyond the level from which spontaneous recuperation is possible.

The issue of conservation recurs repeatedly in these pages, but is neither the theme of the book nor its justification. In the space available, it is equally impossible to offer an identification manual to the species listed in the Checklist. The book's aim is more modest: simply, to illuminate, for all who read it, the wealth of wild mammals in South-East Asia, and to touch on their diversity and their relationships with the natural environment. Many of these mammals have cultural links with folklore and tradition. Those that are legitimately hunted can provide a bountiful source of wild meats. The biology of few is known, and much undoubtedly remains to be learnt in the fields of medicine and zoology. The mammals hold a unique place in the natural heritage of mankind of these lands.

2
Primitive Mammals

THE earliest true mammals evolved from terrestrial reptilian ancestors during the Triassic era, about 200 million years ago. Although plants and animals were totally different, the environment on land at that time was not wholly unlike present-day tropical rain-forest. The climate was warm and humid, and the vegetation a dense jungle of tall, evergreen plants. The first mammals were small, with the sharply pointed cusps of their cheek teeth indicating predatory habits and a diet in which insects predominated.

At that time the dinosaurs were dominant on land, and for the next 100 million years or so, mammals remained small and unobtrusive. Nonetheless, during that immense period of time, several fundamentally different lines evolved. Some persisted; others disappeared. By the middle Cretaceous era (about 100 million years ago) the three major divisions among modern mammals, that is to say, the monotremes, the marsupials and the placentals (see below), had certainly separated. Dinosaurs disappeared towards the end of the Cretaceous era and about 64 million years ago there dawned the Age of Mammals, the Tertiary or Cenozoic era.

Marsupials

Among living types, the monotremes (divergent, egg-laying, toothless mammals) are confined to New Guinea and Australia. The marsupials, or pouched mammals, are also chiefly found in the Australasian region (and South America) but a bandicoot, four species of cuscus and the sugar glider occur on Indonesian islands that are South-East Asian by our present definition. The Seram bandicoot was taken at high elevation on that island. It has been found nowhere else, but its continued existence is unverified. Cuscuses are stocky and close-furred, often differing between the sexes in colour pattern. Their fore-feet have five free, clawed toes, the inner pair opposed to the outer three, while the hind feet have a large opposable inner toe, narrow, conjoined second and third and free fourth and fifth toes. The tail is prehensile, furred at its base, naked and scaly towards the tip. They are arboreal in habit, omnivorous in diet. The females give birth to one or two minute young which immediately transfer to nipples enclosed within the pouch (the 'marsupium'). Here they remain firmly attached for several weeks. Two species occur on Celebes and two others in the South Moluccas and Timur. In many parts of their range, cuscuses are hunted. There must be concern to preserve the Celebes endemics (that is to say, species confined to that one place) and to protect all populations of these interesting marsupials at the western margin of their world distribution.

Placentals

All remaining mammals of South-East Asia belong to the dominant group, the placentals, among which the young are born at a developed stage, having been nourished in the womb via the placenta. Extinct species of late Cretaceous times are known only by their fossilized skeletal remains. The living mammals most closely resembling them in their bony anatomy are the gymnures and treeshrews. For this reason, these groups are judged to be primitive, even 'living fossils', from which we can infer aspects of the soft-tissue anatomy and general biology of the ancestral types. Both groups are now largely confined to continental South-East Asia and the islands of the Sunda Shelf.

Gymnures

Largest among them is the gymnure or moonrat. This is a rat-like mammal, with head and body up to 45 cm long plus a scaly tail up to 25 cm long. It has a woolly underfur, through which grows a ragged coat of coarse guard hairs, grizzled black except for a white mask in moonrats of the Malay peninsula and Sumatra, white overall with a scattering of black hairs in Borneo (Plate 1). The feet have five fully-formed digits terminating in claws, the ears are short and hairless, and the muzzle ends in a pink, flared and mobile snout. The mouth is furnished with the maximum complement of teeth found among placentals, forty-four in total (three incisors, a canine, four premolars and three molars in each row). Moonrats are solitary, active chiefly by night, living on the ground, especially in valleys and damp areas, taking a mixed diet of small terrestrial and aquatic prey such as earthworms, snails, arthropods and frogs. They seem to breed at any time of the year, producing a litter of two young after a gestation period of forty-

1. The moonrat of Borneo, hunched in its typical threat posture.

2. A lesser gymnure.

five to fifty days. Moonrats are famous for their strong ammoniacal stench, originating from glands near the anus. When disturbed, they freeze in an open-mouthed posture, emitting low grunts. This behaviour and their conspicuous colouring together suggest that the foul smell affords sufficient protection against predators.

Other gymnures are short-tailed and smaller. The lesser gymnure (Plate 2) is found in hill forest. It feeds on small invertebrates which it hunts among the litter of dead vegetable matter covering the ground, snuffling under leaves and tossing them aside with quick motions of the head. The Dinagat and Mindanao gymnures are confined to mountains on their eponymous islands; and the shrew-hedgehog ranges through the hilly northern margin of South-East Asia.

Treeshrews

While the gymnures are ground-dwelling, the treeshrews resemble the early ancestors of arboreal types. They occur throughout the western part of the region, reaching greatest diversity in Borneo where ten of the fourteen species occur. Most have squirrel-like bushy tails, grizzled brown or olivaceous coats, marked with streaks or stripes of buff and/or black (Colour Plate 1). Their skulls are conical, muzzles tapered and elongated, and orbits prominent, showing that treeshrews have some degree of binocular vision. Their dentition is slightly reduced from the maximum (to thirty-eight teeth) but the form of the molars, in particular, is primitive. They also retain the primitive foot of five fully developed, clawed digits; the first toes are opposable to some degree. All are arboreal as much as ground dwelling, feeding on a mixed diet of fruit and invertebrate prey. All except one, the pentail, are diurnal. Treeshrews as a rule are solitary or associated in small groups, and apparently territorial. The sexual cycle includes a period of bleeding which has been compared with menstruation among primates. Treeshrews build nests in tree holes or similar sites. Normally two young are born, hairless and with closed eyes, after a gestation period of about forty-five days. They grow rapidly, and leave the nest in little over a month.

Shrews

Of other primitive mammals, the shrews are represented in South-East Asia chiefly by a large number of white-toothed shrews, some musk shrews and a water shrew. The ranges of one red-toothed shrew and the burrowing Chinese short-tailed shrew, essentially members of a more northerly Sino-Himalayan fauna, extend south into the region.

All shrews are small mouse-like mammals with the sharp pointed teeth characteristic of primitive placentals. They are ground dwelling predators of invertebrates of the soil surface and leaf-litter, widespread but difficult to trap by conventional means. Often, capture by zoological collectors has been a matter of chance. When I lived in the Gombak forest reserve, Malaysia, my domestic cat brought in more specimens of the local pigmy musk shrew than existed at that time in all the museums of the world. The entomologist, Mark Collins, achieved similar results by setting small pitfall traps (intended for insects) in the forest floor in Mulu National Park, Sarawak. Elsewhere, without such lucky breaks, collection has been erratic and knowledge of distribution and taxonomy, especially of island forms, is patchy and incomplete.

The pigmy musk shrew, at 2 g body weight, is the smallest of all mammals. The Sunda water shrew is as much at home in streams and rivers as on land, and has a thick waterproof fur, a keeled tail and fringes of stiff hairs at the margins of its feet to assist it when swimming. The house shrew (Plate 3) is a familiar commensal of man. It has been carried by human agency round much of the tropical world, and its original range can no longer be determined.

3. A house shrew.

Moles

The most specialized relatives of these primitive mammals are the moles. While retaining ancient types of tooth and the full placental complement of forty-four teeth, moles show many modifications for their burrowing mode of existence. Thus, the body is fusiform, the snout conical, eyes vestigial, external ears absent, the hair is short and velvety and has no particular direction of lie, the tail is stubby and the limbs highly modified for digging. Principally northern in its distribution, the short-tailed mole has an outlying population in the peaty soils of the high mountains of Peninsular Malaysia.

3
Mammals of the Air

Colugos

AMONG the earliest fossils of the Age of Mammals, found chiefly in North America, are some curious teeth, tending to a triangular shape and furnished with a multitude of small cusps. These resemble the teeth of the flying lemurs or colugos (Plate 4), to which they are thought to be related. If this interpretation is correct the two living species, found only in South-East Asia, are the survivors of a most ancient lineage.

The modern colugos are strange creatures, with sad, gentle

4. Colugo.

faces and soft fur, flecked and mottled black on rusty brown or grey. Their most distinctive feature is a web of skin (a 'patagium') stretching from the neck to enclose all four feet and the tail to its tip. The colugo is strictly arboreal, scarcely able to progress on the ground encumbered by this sheet of loose skin. But once aloft, with the limbs extended, the patagium is stretched and, like an animated kite, the animal can swoop and glide with superb control for great distances from one tree to the next. They are not normally vocal, but I have heard chilling, raucous screams uttered by a solitary adult under attack by a pair of yellow-throated martens.

Colugos give birth to one young at a time. The baby is carried by the mother while it suckles. At the same time she can become pregnant again so that, once weaned, the juvenile is quickly supplanted by a successor. Colugos are chiefly nocturnal but sometimes active in daylight, frequenting forests of all kinds and also venturing into plantations and gardens, especially favouring coconut palms where they find the rich nectar attractive. They subsist off vegetable matter, including soft new shoots, but are not pests. Their flesh is scarcely edible and they are difficult to keep in captivity. They deserve to be protected and cherished as special and most unusual South-East Asian mammals.

The colugos are not the only mammals that can glide. Similar adaptations, involving a furred patagium, have evolved independently in other groups. Marsupial examples in New Guinea and Australia are the gliders of the possum families Burramyidae and Petauridae, one of which, the sugar glider, has been recorded in the North Moluccas. The other, more numerous gliding mammals in South-East Asia are the flying squirrels, of which we will read more later (see Chapter 7). Like colugos, gliding possums and flying squirrels are active by night. It appears that the habit of flight (including gliding) has evolved among mammals as an adaptation to nocturnalism.

This is most emphatically true of the dominant flyers, the bats.

Bats

Bats are superficially so different from other mammals that many people fail to recognize them as such. They are often regarded with distaste, as alarming, dirty and potentially dangerous. Only the Chinese culture, because of the homophony of the characters for 'bat' and 'happiness' (both romanized as 'fu'), looks kindly on these small and intriguingly specialized mammals.

A quick glance will show that the membraneous wing of a bat (a patagium, again) is supported by the skeleton of a five-fingered fore-limb not unlike the human arm and hand with the second to fifth fingers greatly elongated; the first digit (the thumb) is small and largely free of the patagium. The body is generally covered in short fur. The head may be rather dog-like in appearance, but many bats bear complex fleshy outgrowths on the muzzle, above and surrounding the nostrils; these structures are termed 'nose-leaves' (Figure 1). The bat may or may not have a tail, and this may or may not be involved in the patagium joining the hind limbs. While undoubtedly of biological importance, such details are also useful features figuring in all published keys to the identification of bats.

The bats of South-East Asia are numerous and diverse, inferior only to rodents (Chapter 7) in the number of species, amounting to 32 per cent of all mammals. The first bats appeared early in the Age of Mammals already fully adapted for flight and differentiated into the two main groups, Megachiroptera and Microchiroptera. The fossil record gives no clue to their ancestry, but features of the teeth of Microchiroptera resemble those of the primitive insectivores.

Their ability to fly has allowed bats to penetrate to distant

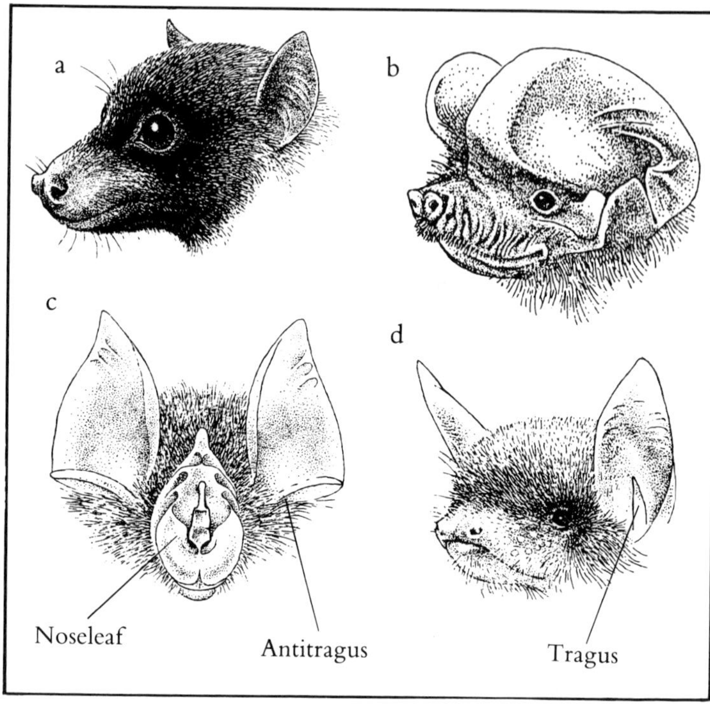

Figure 1. Heads of bats (not to scale). (a) A fruit bat, (b) wrinkle-lipped bat, (c) horseshoe bat, showing noseleaf, and (d) vespertilionid bat.

quarters of the habitable globe, reaching the limits of five continents and many remote islands. Yet, in South-East Asia, the species of bats are not uniformly distributed but fall broadly into three subregional assemblages occupying respectively the continental mainland, the Sunda Shelf and the islands further east. The bat faunas of the Philippines and Celebes are distinctive, showing some mutual affinity. Many species have ranges overlapping these divisions and the borders between them cannot be sharply drawn. Still, analysed in this way the zoogeography of bats is a model for all mammals of South-East Asia, reflecting the influence of continental Asia as the chief source region, the importance of areas of local speciation on islands of the Sunda Shelf, the Philippines and Celebes, and finally the influence of New Guinea as a secondary source region in the east.

The Megachiroptera are commonly called 'fruit bats' and the Microchiroptera 'insectivorous' bats, respectively. Beyond the major difference in diet implied by these names–vegetarian versus predatory hunters–there are other important distinctions. Chief among these is the method of orientation. Fruit bats have prominent nostrils and big eyes. The

5. Head of the horseshoe bat *Rhinolophus luctus*.

corresponding brain centres for smell and sight are well developed, and these senses are clearly of great importance in their lives. The insectivorous bats do not show similar development of the central nervous system, and in fact orient by a sophisticated process of echo-location, based on very high frequency vocalizations. The nose-leaves of those bats that have them (Plate 5) serve to focus and direct the beam of emitted sound. The bat's proportionally large external ears, with a prominent tragus or antitragus (see Figure 1), emphasize the importance of hearing. Hard though it may be for us to understand, it is evident that the picture of its surroundings detected by a bat by means of echo-location is as complete as ours obtained by vision. Indeed, long ago, it was shown (in an experiment we would today consider cruel) that surgically blinded insectivorous bats can continue to fly, to catch their prey and apparently to pursue a normal life.

Among fruit bats only the rousettes, as far as known, possess a limited capacity to echo-locate. In this case, the sound is of lower frequency, audible to man as an interrupted buzz, and is created by action of the tongue (rather than the larynx). The echo-locating system of rousettes is adequate to find their way in flight in the darkness of a cave but, since they are fruit-eaters, it has no function in detecting prey.

Megachiroptera

Megachiroptera, as the name implies ('mega' meaning big) are generally larger than Microchiroptera ('micro' meaning small). Largest of all are the flying foxes; the wing-span of some attains 1.5 m and body weight up to 1.5 kg. Flying foxes occur throughout the Indo-Pacific region. Their range includes islands close to the eastern coast of Africa, but for some reason none has established a foothold on the African mainland. Among the most widespread in South-East Asia is the large flying fox (*Pteropus vampyrus*). These bats are gregarious, gathering at roost in trees in noisy, active crowds. Established sites may be used for years, as in the presidential gardens at Bogor, Indonesia, on the bank of the Tembeling below Taman Negara, Malaysia, and on Siarau Island, Brunei. The bats hang head downwards from the branches except while defecating when they revert to a head-up position, suspended from the thumbs. In the heat of the day they are seldom still for long. Individuals will flap their wings, or take flight and circle before returning to roost. The many interactions between neighbours are fascinating to watch. Shortly before dusk, they begin to depart, forming a long, straggling flight-stream headed towards their nocturnal feeding grounds.

South-East Asia is an important area of diversification of these bats. Few species are widespread; only Geoffroy's rousette occurs across the entire region. One assemblage, including the grey, spotted-winged, black-capped, Dayak, tail-less, and dusky fruit bats, is principally Sundaic in distribution; another, notably the bare-backed and short-nosed fruit bats, is centred in New Guinea. Several, including many *Pteropus* species, all *Acerodon*, the Katanglad, Mindoro, Luzon and Philippine fruit bats, and the Manado, small-toothed and stripe-faced flying foxes are confined to one or a few islands among the Philippines, Celebes group, Moluccas or Lesser Sundas. Records are based on old collections, and in many instances details of present distribution and numbers are lacking.

Most of these bats are forest-dwelling. Their diets are broadly similar, consisting of soft fruits and floral parts, mostly of forest plants (especially wild figs) although a few bats also come to cultivated fruit trees, notably the large flying fox and the dog-faced fruit bats. Characteristically, small fruit are plucked and carried in the mouth to a feeding roost. Here the bat eats the fruit, if it is large taking several bites while steadying it with a foot. The mouthful is usually sucked and squeezed between tongue and palate, and the pulp ultimately ejected as a compact pellet. Small seeds are contained in the pellet, while larger pips or stones are simply dropped. In both cases, it is clear that the bats are effective agents of dispersal. In this capacity, they must play an essential role in the total forest ecology.

The Macroglossinae are specialized nectar feeders. They have narrow, tapering muzzles and long tongues with brush-like tips. The blooms of some forest plants display an array of characters designed to attract these bats, which are presumed to serve as pollinators. Typical features of 'chiropterophilous' (meaning bat-loving) flowers include an exposed position on branches or at the ends of twigs, nocturnal opening, strong scent, copious nectar and a pale, dull colour, such as cream or light green. Familiar examples are the durian (*Durio zibethinus*), petai (*Parkia speciosa*), wild kapok (*Bombax*) and bananas (Plate 6).

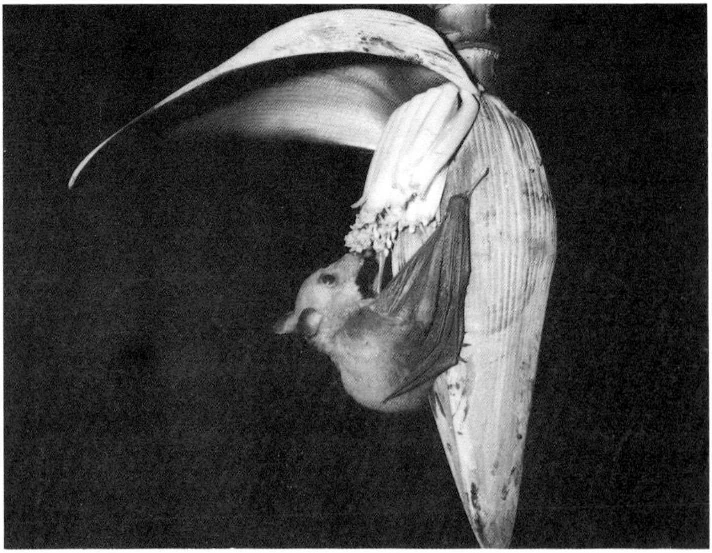

6. The smaller long-tongued fruit bat *Macroglossus minimus* takes nectar from a banana flower, photographed at night.

The bats settle in order to probe for nectar and, while doing so, they become dusted with pollen. It has been shown that the cave fruit bats of Batu Caves, Peninsular Malaysia, visit durians in great number during the brief flowering season, but rely on wild trees (notably the mangrove tree, *Sonneratia*) to provide basic year-round sustenance. The long-term survival of a pollinator of an important commercial fruit thus depends on the continued presence of a natural plant community.

Microchiroptera

Among Microchiroptera, the large false vampire is flesh-eating, preying on small bats, birds (taken at roost or from the nest), geckos and frogs. Other microchiropteran bats feed on arthropods, principally insects, in some cases gleaned from vegetation (for example, the lesser false vampire) but mostly caught in flight. Bats also drink on the wing, dipping down over water to sip from the surface.

The eight South-East Asian families are distinguished by anatomical differences including the form of the shoulder joint, aerodynamic characteristics of the wing and structure of the tail. These features are reflected in the mode of flight which is, for instance, fast and direct among the Molossidae, slower and more erratic among the Rhinolophidae. As an organ of flight, the bat's patagium is much more controlled and versatile than the feathered wing of a bird. It can also have important subsidiary functions, for instance, in the capture of flying prey, in the processes of thermoregulation and even holding the emerging baby bat at its birth.

The roosting habits of these bats are immensely varied. Some are gregarious, forming colonies of tens of thousands such as the wrinkle-lipped bats which emerge in spectacular evening flights from the Deer Cave, Mulu National Park, Sarawak, or Gomantong Caves, Sabah. Many favour caves (Plate 7), others

7. Leaf-nosed bats *Hipposideros diadema* at roost in Niah Cave, Sarawak.

tree hollows, bunches of dried leaves or palm fronds. Several, especially the white-bellied tomb bat and the house bats, commonly occupy roof spaces, often to the annoyance of the householder.

In some cases different preferences apparently define the special niches of closely related species, thus allowing them to coexist. Among mouse-eared bats, *Myotis horsfieldii* roosts in caves and rock crevices while *M. muricola* chooses the furled central leaf of banana plants. Between the two flat-headed bats, both of which often occupy the internodal spaces of large bamboos, the distinction is more subtle, being based chiefly on the width of the slit-shaped entrance holes (Plate 8). The larger flat-headed bat has been found where the aperture ranged from 4.6 to 8.7 mm wide, and the smaller from 6.6 mm down to 3.9 mm. No mixed roosting groups are known, but both species have been found on different occasions within nodes entered by holes in the width range 4.6–6.6 mm. The lower limit is

8. A flat-headed bat emerging from its roost within the internodal space of a large bamboo; the band on the forearm was applied during a research programme.

probably dictated by the bat's skull size, but the upper must reflect an act of choice.

Most of these bats give birth to one young at a time; some Vespertilionidae (including the flat-headed bats and the house bats) normally bear twins. Many have regular breeding cycles, even at low equatorial latitudes where there is negligible seasonal variation in day-length, temperature or rainfall. At Ulu Gombak, Malaysia, just above 3°N latitude, the greatest spread of birth dates of the flat-headed bats was twenty days, and the median date varied by only one week in two different years.

In detail, in the males of many species there is a distinct reproductive cycle which culminates in total exhaustion of the germinal epithelium within the testis and engorgement of the epididymis with sperm. The mating season may then be delayed somewhat, so that fertile copulations depend entirely on this store of sperm. In some Rhinolophidae and Vespertilionidae there is a further pause after insemination, a period during which the sperm is held in the female tract, sometimes for weeks, until ovulation occurs. Among the South-East Asian bats, these mechanisms are known to occur even in populations at low equatorial latitudes where the advantage of co-ordinated seasonal breeding is not immediately obvious. The elucidation of the successive triggers to each stage in the cycle poses a major problem for research. Moreover, the physiological processes, whereby the female system becomes temporarily tolerant of the foreign (male) cells, are of great medical interest, particularly in the field of human fertility.

COLOUR PLATES

a

b

c

d

e

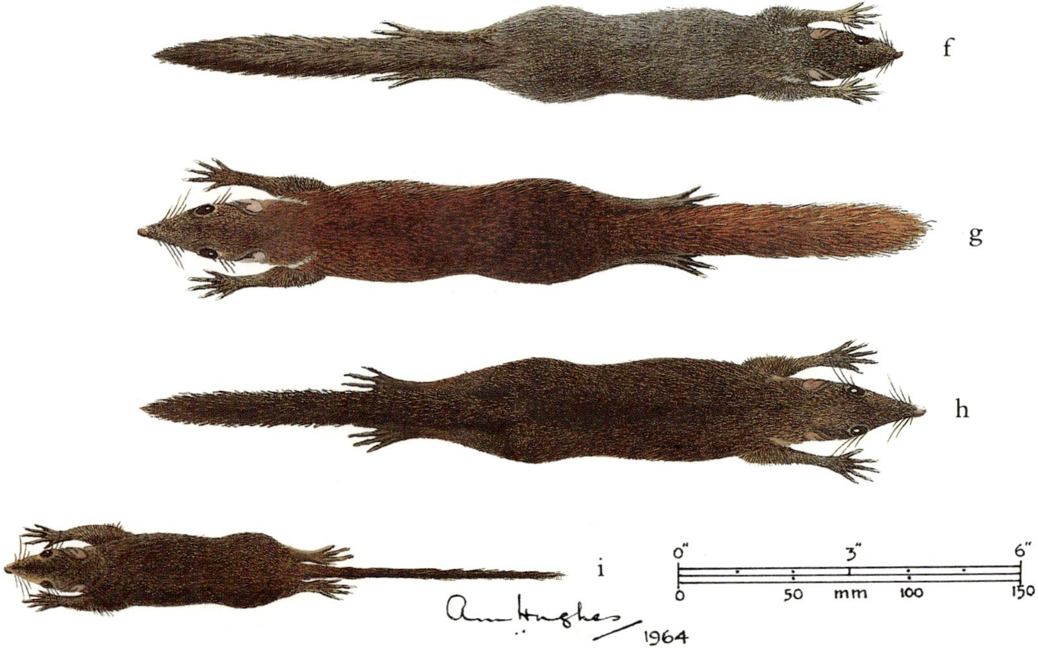

1. Treeshrews. A selection of South-East Asian species, illustrated in stylized posture to show the differences in coat colour. (a) Lesser treeshrew. (b) Painted treeshrew. (c) Large treeshrew. (d) Common treeshrew. (e) Striped treeshrew. (f) Slender treeshrew. (g) Ruddy treeshrew. (h) Mountain treeshrew. (i) Bornean smooth-tailed treeshrew.

a
b
c
d
0" 6" 12" 18" 24"
0 200 mm 400 600

2. Carnivora: cats and mustelids. (a) Bay cat. (b) Marbled cat. (c) Flat-headed cat. (d) Leopard cat. (e) Smooth otter. (f) Small-clawed otter. (g) Ferret badger. (h) Teledu.

a
b
c
d
0"
6"
12"
0
100
mm
200
300

3. Carnivora: civets. (a) Banded civet (*Hemigalus derbyanus*). (b) Malay civet. (c) Banded linsang. (d) The Bornean form of the common palm civet (in which the spots on the back are very obscure). (e) Masked palm civet (the variant with the white tail tip). (f) Three-striped palm civet. (g) Hose's civet. (h) Otter civet.

a
b
c

4. Colourful squirrels. (a) Kinabalu squirrel. (b) *Callosciurus prevostii pluto*, a variety of Prevost's squirrel found in the north of Borneo. (c) *C. p. caroli*, a variety found in Sarawak from the north bank of the Rejang River to Limbang and Lawas districts. (d) *C. p. borneoensis*, a variety found from the Landak River, West Kalimantan, to the Saribas, Sarawak. (e) *C. p. atricapillus*, a variety found across the centre of Borneo island limited on the north by the Rejang River, Sarawak, and the Mahakam River, East Kalimantan.

a
b
c
d
e
f

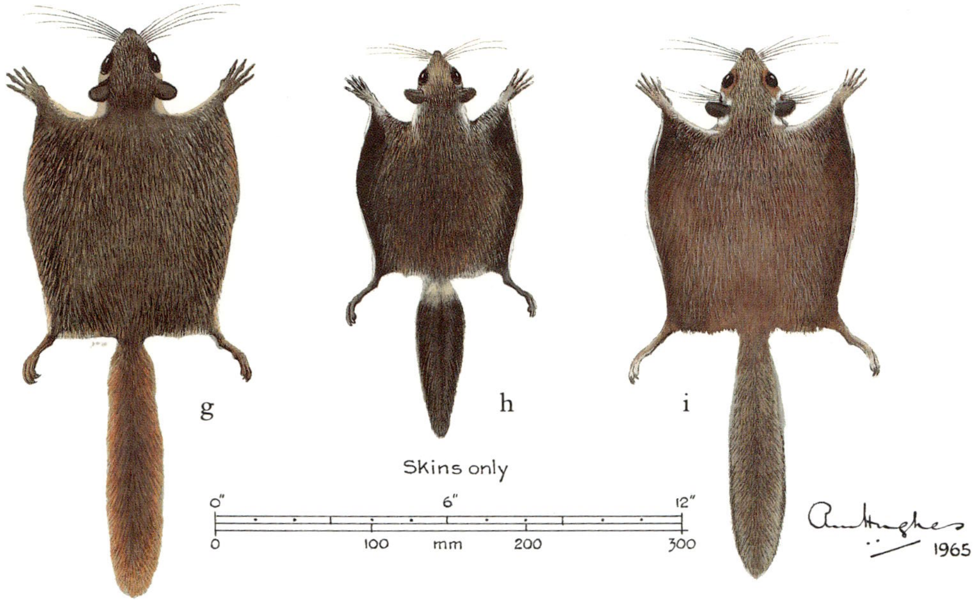

5. The smaller flying squirrels. (a) Horsfield's flying squirrel. (b) Pigmy flying squirrel (*Petaurillus emiliae*). (c) Whiskered flying squirrel. (d) Hose's pigmy flying squirrel. (e) Red-cheeked flying squirrel. (f) White-bellied flying squirrel. (g) Horsfield's flying squirrel. (h) Grey-cheeked flying squirrel. (i) Whiskered flying squirrel.

a
b
c
d

6. Rats. (a) Red spiny rat (*Maxomys surifer*). (b) Long-tailed giant rat. (c) Mountain giant rat. (d) Summit rat. (e) Mueller's giant rat. (f) White-bellied rat (*Niviventer cremoriventer*). (g) Whitehead's spiny rat (*Maxomys whiteheadi*), normally with a bicoloured tail, but dark-tailed in the population on Miang Besar Island, East Kalimantan, providing a link with *Maxomys musschenbroeki* of Celebes.

4
Primates

THE primates originated early in the Age of Mammals from ancestors resembling treeshrews. General features are the opposable first digits of hand and foot, and forward-looking eyes providing binocular vision and hence the ability to judge distances. Both features are adaptations to an arboreal life and indeed all South-East Asian primates are forest-dwelling.

Tarsiers

Smallest of South-East Asian primates are the tarsiers, weighing about 100 g (Plate 9). Close relatives occurred in North America and Europe early in the Age of Mammals, but the three living species are now confined to South-East Asian islands. Little is known in detail of their present numbers and distribution, but tarsiers are generally not hunted–indeed, in some parts, they are regarded as animals of ill omen.

The fur of tarsiers is thin and soft. The digits end in expanded pads, with small, flat nails on all except the second and third of the foot which carry raised claws. The first toe is strongly opposable, the thumb less so, but both hand and foot can grip firmly. The tail serves as a prop, but is not prehensile; in the naked mid-portion, the skin is finely creased but not scaly. Tarsiers are nocturnal, as attested by their huge, forwardly directed eyes. The orbits are separated only by a thin sheet of bone. They inhabit the undergrowth of thick forest, leaping from perch to perch by powerful thrusts of the hind limbs. They hunt large invertebrates and small vertebrates such as geckos. Males apparently occupy largely exclusive territories, which may overlap with the ranges of more than one female. Females

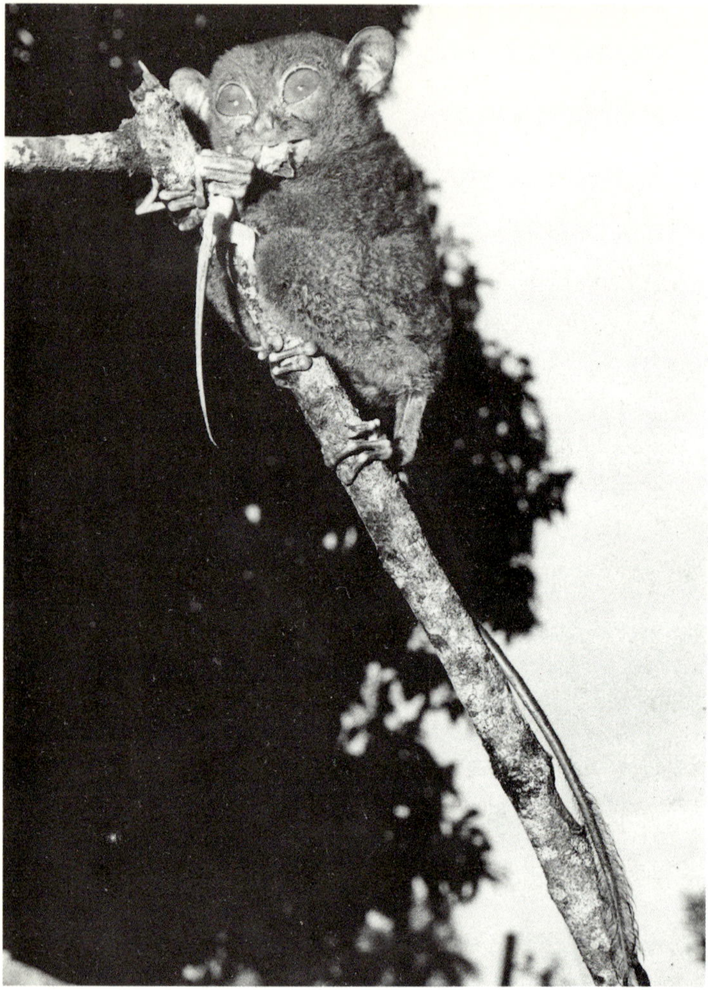

9. A western tarsier *Tarsius bancanus*, photographed in sunlight.

undergo a sexual cycle; swelling of the external genitalia and uterine bleeding accompany the heat. Only one young is born; it is fully furred and active from birth.

Lorises

The lorises (Plate 10), somewhat larger (0.5–0.9 kg weight), are also nocturnal. They forage in the forest canopy, clambering with slow deliberation, never releasing their grip. The tail is vestigial. The thumb of the hand is rudimentary, but the first toe of the foot is large and opposable. All digits bear flattened

10. Slow loris *Nycticebus coucang.*

nails except the second of the hind foot, which carries a large, curved claw, said to be used in grooming. In the wild, they are normally solitary or associated in pairs or in parent–offspring couples. Females undergo sexual cycles, during which the external genitalia become swollen. At about six and a half months, the gestation period is long for a mammal of this size. One young is born, fully furred, and immediately clings to its mother.

Monkeys

The monkeys belong to a family (Cercopithecidae) which occurs across the tropics of Africa and Asia. South-East Asia has been an important centre of local evolution: eleven of the twenty-four species occur on single islands or small island groups. There is a tradition of training wild-caught pig-tailed macaques to pluck coconuts. Some no doubt escape, or are released when they become intractable, and thus the range of this monkey may have been artificially extended by man. Long-tailed macaques (Plate 11) are also easily tamed, and as pets are carried beyond their natural range. The most widespread unaided distribution is that of the silvered langur. This monkey is successful in many habitats, including coastal vegetation, and retains the potential to expand its range naturally. Thus, while it occurs far inland in Sumatra and Borneo, in Peninsular Malaysia it is confined to a strip no more than 40 km deep on the west coast, apparently a relatively new invader from Sumatra. On the other hand, although it is a mangrove specialist, the proboscis monkey of Borneo remains confined to that island. On Celebes, topographical peculiarities and recent geological history are believed to be responsible for the radiation of local macaque stock into four distinct species. The pig-tailed langur (or 'simakobu') and Mentawai langur are confined to small islands west of Sumatra; the problems of their conservation attract international attention.

11. Adult female long-tailed macaque.

The monkeys have coarse, often shaggy coats, generally drab olivaceous brown among macaques, but more colourful among the langurs–rusty red, brown, grey, black or white, plain or patterned. The young of many species are coloured differently from the adults, generally being more conspicuous: for instance, orange in the grey adult-coated dusky and silvered langurs. The facial skin may be brown (for example, in the long-tailed macaque), brown suffused with red (stump-tailed macaque), black (Celebes macaque, *M. nigra*), black with white markings (white-fronted langur, dusky langur) or blue (maroon langur). The digits are long, and all end in curved nails. The thumb tends to be foreshortened among langurs, but the first toe is strong and widely opposed in all. All arboreal species have long tails, important in keeping balance but not prehensile. The tail is shortened among ground-dwelling species, notably the pig-tailed, stump-tailed and Celebes macaques and pig-tailed langur.

All monkeys are diurnal, retiring before nightfall each day to roost in trees, where they remain until dawn. Excepting some small island populations, they normally associate in groups consisting of more than one adult male, a greater number of adult females and still greater number of subadults and juveniles. These large groups may separate temporarily into subgroups, of which the adult female and her offspring form the minimum unit. Group territories are defended against members of the same species, although the presence of others is generally tolerated. All the monkeys have characteristic loud calls, and vocalization is important in communication within and between social groups.

The macaques take opportunistic, omnivorous diets, including animal prey, fruits and leaf shoots. The langurs feed on a higher proportion of leaves and young shoots. They possess large, sacculated stomachs, in which bacterial fermentation assists the breakdown of plant tissue, including cellulose. The proboscis monkey is unique in its capacity to digest the leathery leaves of *Sonneratia* trees of the mangroves, a principal component of its diet.

Like the tarsiers and lorises, female monkeys undergo regular sexual cycles, during which the external genitalia swell. One young is born, fully furred and active, and is carried by its mother. While parental ties are strong, and the mother will remain close to her offspring long after it is weaned, young monkeys also indulge in multiple social contacts with other adults in the troop and extended bouts of play with those of their own age.

Monkey-watching is fun. Long-tailed macaques are most

easily found, sometimes too common (even pestilential) in town areas or reserves such as Bako National Park, Sarawak. Silvered langurs are tame and approachable at Kuala Selangor, Peninsular Malaysia. In remoter rural areas, monkeys are all too often persecuted, hunted for food or harried as pests of tree-crops, orchards or paddy-fields.

Gibbons

The gibbons (Plate 12) occur through South-East Asia west of the Makassar Straits. They lack tails and are long-limbed, with rather short first digits and long, curved second to fifth digits. Their fur is dense, variously coloured black, black and white, grey, brown or honey coloured, with black facial skin and

12. A grey Borneo gibbon, one of several colour variants of this species.

gentle, pleading expressions somewhat belied by long, dagger-sharp canines readily exposed in threat. Apart from the large siamang (about 10 kg weight), all gibbon species are rather close in size (5–6 kg). The different species occupy mutually exclusive ranges (Figure 2), overlapped only by siamang. The boundaries between ranges follow natural features, notably the lower courses of large rivers. In the headwaters, some mixing may occur but natural wild hybrids have rarely been seen.

The normal social group among gibbons is the mated pair plus one or more (rarely in excess of three) immature offspring. These family parties are permanently resident in territories of some 15–60 ha in extent, depending on the quality of the habitat. Territories are fiercely defended against members of the same species (but not against siamang in areas of overlap). Territorial boundaries are defined by occasional confrontations, involving chases and even fights, and reinforced by bouts of calling, usually early in the day, during which the mated pair co-ordinate their cries to reach a crescendo in which the female utters a resounding 'great call', characteristic of each species. Among siamang, a throat pouch fills with air and is distended while calling, enhancing the resonance of the sound.

The gestation period is seven to eight months. One young is born, alert and fully furred and, as among other primates, is carried while small by its mother. The interval between births is normally a couple of years, and offspring remain with the parental group until reaching maturity at about six years.

Orang-utan

The only great ape of Asia, the orang-utan, is found in forests in parts of Sumatra and Borneo. Man and ape have been in conflict for many years, and the future conservation of orang-utans is of international concern. Rehabilitation stations have been set up (for example, at Ketambe in Sumatra and at Sepilok

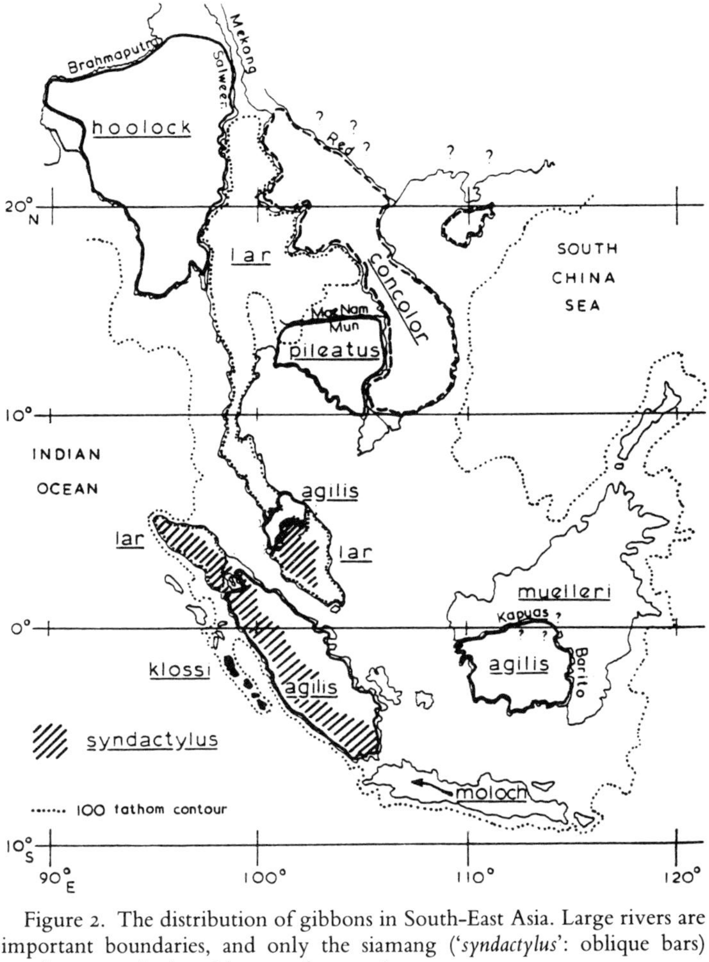

Figure 2. The distribution of gibbons in South-East Asia. Large rivers are important boundaries, and only the siamang ('*syndactylus*': oblique bars) overlaps extensively with any other species.

in Sabah) where former captives are released into the wild. At such places, it is easy for the visitor to see free-living orang-utans, but greater emphasis is now placed on the establishment of secure reserves for natural populations.

Although they are large (adult females 40–50 kg and adult males 60–90 kg), orang-utans are the most arboreal of the great apes. Only the largest males spend long periods of time on the ground. The diet includes foliage, fruit and a small amount of animal food. Normally, before sleeping, an orang-utan will pull together branches in the tree crown to make a rough platform or 'nest' on which to rest. Many surveys, from ground and from the air, have been based on counts of these nests.

Individuals are generally solitary, occupying habitual home ranges and advertising their presence by loud calls, but not so aggressively defending the territory as gibbons. The range of an adult male will overlap with those of several females. Average densities among natural wild populations vary from one to five individuals per sq km. Only during the period of mating will a male and a female remain consistently together. The gestation period is eight to nine months, and the young remains close to its mother (becoming increasingly independent) until about seven years old.

Conservation

As a group, the primates occupy a niche that is unique in the South-East Asian forest. Their ability to climb and move freely and rapidly through the tree crowns allows them to exploit the primary production of the forest canopy. In this role the monkeys, gibbons and orang-utan dominate all other mammals, and in the natural state can be likened to the herds of hoofed mammals cropping the primary production of tropical savannahs. Few detailed observations have been attempted, other than in the lowland forest of Kerau Game Reserve, Peninsular Malaysia. Here, a synthesis of results of several years of study by different workers indicates a generalized average sum of about 180 higher primates (four kinds of monkeys, white-handed gibbon and siamang) per sq km, equivalent to an aggregate mass of about 835 kg per sq km. This figure is far lower than comparable values for savannah grassland in equatorial Africa, where even single species may amount to tons: for example, hippopotamus 11.2 tons and African buffalo 6.1 tons per sq km in the Rwindi–Rutshuru plain, Congo. The discrepancy emphasizes a major problem confronting conservationists, namely, the huge areas of good quality forest that need to be set aside–in the face of all other conflicting demands–if adequate populations of these unique South-East Asian primates are to be safeguarded and preserved for posterity.

5
Carnivora

THE main predator of monkeys is man, who seeks them for their flesh or merely for the bezoar stones in their bellies (possessed by about one in seventy, hunters have reported), destroys them as pests or captures them for the laboratory trade. Besides these depradations, the toll taken by carnivorous wild mammals is trivial. Indeed, the tree-tops seem to be a safe refuge since even those large carnivores able to climb mainly seek their prey on the ground.

Wild Dogs

Like all members of the dog family, the jackal and dhole hunt only ground prey. The jackal occurs in Africa and across all Asia, reaching its eastern limits in Thailand. It is an animal of the forest edge and open country, often scavenging around human habitation. The dhole also occupies an extensive range in central and eastern Asia and, being a true forest dog, has been able to penetrate to the limits of the Sunda Shelf. Although now absent, archaeological evidence suggests that it was in Borneo some 10,000 years ago.

Dholes hunt in pairs or small packs, probably family groups. Their prey includes insects, small vertebrates and mammals larger than themselves, such as sambar and barking deer. These are run down, and thrown by pack members biting and holding the head. After the kill, the dhole's first act is to open the belly. The abdominal skin and body wall are cut by the scissoring action of the last upper premolar slicing against the first molar in the lower jaw. These teeth are called the 'carnassial' pair and are characteristic of the Carnivora. Choice items of the viscera such as the liver are first consumed, although some of the entrails may be spurned. After it is lightened and partially dismembered, the carcass (or bits of it) will be dragged into cover. Ultimately, all the flesh will be consumed, and bones, skin and other rejected parts left scattered.

It is inevitable that a predator must be less abundant than its prey. Although widespread, dholes are rare, and only met by chance. From studies in zoos, it is known that the gestation period is about ten weeks and, like domestic dogs, the pups are born helpless and with closed eyes. They are whelped in a den and, as they begin to be weaned, at first are fed chiefly on vomited food.

Bears

Of the two bears, the black bear ranges across central and eastern Asia, south in Thailand at least to Khao Yai National Park where I once had a close encounter. Bear and man were equally startled! The sun bear reaches into Assam, but is otherwise exclusively South-East Asian.

These bears are forest dwelling, widespread but nowhere common, ranging from the lowlands to mountain summits. Both can climb with agility, supported by their powerful claws. Sun bears are partial to durians, and many a straight and lofty durian trunk is scarred by the gouged traces of bears' claws. Fallen durian fruit are opened on the spot, the flesh sucked off and the seeds ejected at the same place. If your durian orchard is pilfered, a neat pile of claw-marked husk segments and seeds is evidence that a bear was the culprit; men are not so tidy. Bears are also fond of honey and bee grubs, and their powerful claws equip them to tear open hollow trees to obtain these foods, again leaving characteristic scratches.

Martens and Weasels

Among Mustelidae, the yellow-throated marten, which is widespread in the Himalayas and eastern as well as South-East Asia, is a truly arboreal small carnivore. Family parties of three to six can be seen gambolling together in the high branches of tall rain-forest trees, and I once came across a couple attacking a solitary colugo as it tried to climb a vertical trunk. The martens were able to cling to the tree by their claws, darting up and around their quarry, nipping its feet and patagium. Their normal diet also includes smaller prey, such as lizards, nestling birds and large insects.

The weasels, equally carnivorous, are ground-dwelling. The Siberian weasel, with a huge range across eastern Asia, and the Himalayan striped weasel both extend into the open country of northern Thailand, feeding chiefly on ground rodents. The Malay weasel occupies the humid rain-forests of the Sunda region. The Javan weasel is most closely related to the Siberian, providing one of several examples of an interrupted distribution linking Java, experiencing a comparatively seasonal climate, with the monsoonal zone of continental South-East Asia.

Badgers

The marten and weasels have a distinctive sharp odour, originating from scent glands at the anus. Among the South-East Asian badgers, these odoriferous glands are especially well developed and the secretions they produce are notoriously foul-smelling. The largest of the group is the hog badger (7–14 kg), smallest are the ferret badgers (1–3 kg). All have rather coarse, dull brown or brownish-grey fur, with prominent white or white and black linear markings on the head and/or back. The most contrasting pattern is that of the teledu (Colour Plate 2), and it is this mammal which produces the foulest-smelling secretion. As one witness has described, when molested, the teledu raises its tail straight up in the air, turns its head away and ejects to a distance of 'some six inches or more' (> 15 cm) nearly a teaspoonful of pale greenish fluid, the smell of which is almost enough to make a man sick and, reputedly, sometimes will asphyxiate a dog or blind it, if struck in the eye. The black and white pattern of the teledu's coat is clearly an example of warning coloration, as in the case of the equally notorious American skunk.

All the badgers dig deep burrows into which they retreat by day, emerging at night. All have strong feet and large claws, and prominent pink snouts. The ferret-badger will climb, but the others are ground-dwelling. All feed on a mix of roots, tubers, soil-dwelling animals and other invertebrates and small vertebrates, if encountered. The young are born in the burrow. Litter size is normally two to three young.

Otters

The otters (Colour Plate 2) are aquatic, and show many adaptations for their way of life, such as dense, water-repellant fur, partially webbed feet and powerful, often flattened tail to provide the propulsive drive in water. Smallest is the small-clawed otter (3 kg), which is also distinguished by its round-ended, finger-like toes, with only vestigial claws. This little otter has a most delicate sense of touch and, when foraging, spends much time groping in soft stream-bottom sand or mud, or feeling among stones or in crevices under water, from time to time bringing up its catch for inspection and (if acceptable) for consumption. The *Lutra* otters are larger (the smooth-coated otter weighs up to 11 kg) and have webbed but otherwise more dog-like paws and digits with normal claws.

The small-clawed otter has been attributed with an exclusively freshwater range, but evidence suggests that all the

South-East Asian otters are equally at home in the sea, rivers or small inland forest streams. They also travel freely on land. They have broad diets, including fish, crustaceans, molluscs and aquatic insects and, no doubt, amphibians and other prey taken on land. The small-clawed otter is most gregarious, associating in parties of up to a dozen; the *Lutra* species less so, up to six. They communicate by shrill, yelping calls and whistles.

Civets

The civets (Viverridae) form a most interesting group, richly represented in South-East Asia, where they are adapted to widely varying modes of life. They are mostly dull coloured, light or dark brown, marked with bars, spots or stripes, and frequently with a dark or pale facial 'mask' and ringed tails (Colour Plate 3). In size, they vary from the binturong (up to 20 kg) to the banded linsang or small Indian civet (about 2 kg). The binturong is fully arboreal, sleeping by day in the tree-tops, and furnished with sharp, curved claws and a prehensile tail to facilitate climbing. While bears slide down tree trunks in a head-upwards posture, the binturong slithers head first, most rapidly. The linsangs and palm civets also climb with ease and spend much of their time aloft, sleeping by day and foraging by night. The common palm civet is often found in the roof-spaces of rural houses and for long has probably had an association with man. Intentionally or inadvertently, this civet has been widely spread among the more easterly islands. The *Viverra* species and banded civets are less scansorial, Hose's civet is largely ground-dwelling and the otter civet is riparian or aquatic.

The diets of civets are individually variable and opportunistic. Most will take fruits, and in forests it is commonplace to find packets of germinating seeds in old droppings, witnessing their importance as dispersal agents. Indeed, the extra tang of the notorious *kopi luak* (civet coffee) of Java purports to derive from this particular selection process! Most civets also take animal prey, when the chance arises, including small vertebrates and invertebrates.

The civets have a nearly full complement of teeth, in these South-East Asian species comprising three incisors, the canine, four premolars and one or two molars in each jaw (a total of thirty-eight or, mostly, forty teeth). The dental morphology, however, is variable and can be correlated with feeding behaviour. Thus, linsangs have sharp-cusped teeth and well developed blade-like carnassials, implying predatory habits and a diet of flesh. In the banded civets, the teeth have many sharp, conical cusps, resembling those of insectivores such as the moonrat, and indicating a similar mixed insectivorous diet. By contrast, binturong and the palm civets have low rounded cusps to the teeth, the carnassial pair lacking their characteristic shearing edges. These features suggest a diet in which fruits and other soft foods predominate.

The rare and peculiar otter civet (Colour Plate 3) possesses premolars that are large and triangular in profile, while its molars are low and broad. In these features, its teeth resemble those of fish-eating mammals, presumably for the same purpose–the premolars to capture and hold wet and slippery prey, the molars to crush and chew. The recorded diet of this civet includes fish, crabs and freshwater molluscs, with other prey. In its thick, water repellant fur and partially webbed feet, it shows typical aquatic adaptations. However, its short tail is clearly of no use in powering its swim. The strange long, stiff white whiskers must have a tactile function. It has been suggested that the otter civet's mode of hunting is to lurk in ambush at the water's edge, taking advantage of its upwardly pointing nostrils to lie largely submerged, and to pounce on any small creature passing by.

Also unusual in its habits is Hose's civet (Colour Plate 3). This seems to be mainly terrestrial. It has an expanded, pink tip to its nose, reminiscent of the moonrat's, and presumably serving to search and snuffle out earthworms and other invertebrates in the peaty soil and leaf litter of the montane habitat favoured by this rare Bornean animal.

Little is known of the breeding habits of civets. Evidence from captive individuals suggests that females first come into heat a little under two years of age, and thereafter undergo biannual cycles. Reported gestation periods are forty-five days for small-toothed palm civet, three months for binturong. Litter sizes are small, in the range of one to three young among South-East Asian examples. A nest is made by the female, into which the babies are born, furred but helpless, with closed eyes. Life spans in captivity range up to twenty-two years for civets of all sizes.

Mongooses

Hose's mongoose, of Borneo, is an enigma. The name was given to a specimen externally indistinguishable from the short-tailed mongoose but differing in cranial anatomy, notably in the shape of the lower jaw. No subsequent example has been obtained and the individual may have been a freak. The other, more familiar mongooses are all active predators. The smaller species are reputed to be willing to tackle vertebrate prey as big as or bigger than themselves, including fowls in the hen-house and cobras! However, rats, frogs, lizards and large insects are more usual. The crab-eating mongoose is semi-aquatic, and takes crabs, frogs, fish and molluscs. Reported gestation periods are six weeks for the Javan mongoose, nine weeks for the larger species; litter size is typically two to four young.

Cats

The claws of mongooses are long and non-retractile. Those of some civets are partially (for instance, *Viverra* species) or fully (the linsangs) retractile. Among the last family of Carnivora, the cats, retractile claws are fully developed. The mechanism involves over-extension, causing backward flexure of the terminal joint of the digit. Cats also have a reduced dentition, consisting of three incisors, the large canine, which delivers the killing bite, two or three premolars in the upper jaw, two in the lower and one molar in each jaw (total twenty-eight or thirty teeth). The teeth are sharp and the carnassial pair large and blade-like. The small number of teeth permits foreshortening of the muzzle, so that the eyes can be forwardly directed, giving binocular vision, essential to these predatory hunters which pounce on or run down their prey. The short muzzle also increases the leverage of the jaws.

The largest cat, the tiger, is threatened throughout its range. That the tiger was formerly also in Borneo is suggested by the discovery of a single canine of a juvenile in archaeological deposits at Niah Cave, at a level associated with Neolithic cultural remains. It is improbable that man of that time would have been able to transport and keep a pet cub. At 150–250 kg, the adult tiger can be a problem to any keeper!

The adult female tiger with dependent cubs forms the basic social unit. Females are normally territorial while the home range of a male will overlap several females. Tigers are solitary hunters and will return to a kill repeatedly until all is eaten. In the South-East Asian rain-forests, wild pigs are their principal quarry; deer and smaller prey will also be taken. Cattle are at risk in rural areas and jungle people, such as the Orang Asli of Peninsular Malaysia, occupying flimsy houses in remote villages, live in permanent anxiety for their own safety. Tigers can develop man-eating habits by raiding the shallow graves of these people. With this background, there are many obvious conflicts to be resolved by the conservationist.

While a tiger's weight prevents it from hunting arboreal prey, leopards (45–60 kg) can climb with ease. Yet it appears that they too use trees only as vantage points from which to drop on prey (including monkeys) and safe retreats to which to carry their catch. The proportion of black leopards increases towards the south, and in Peninsular Malaysia this colour phase outnumbers the spotted. The clouded leopard is half as small again (16–23 kg), and even more arboreal. Yet its weight will still prevent free movement through the forest canopy, and the clouded leopard, too, probably makes most captures by pouncing from a perch.

The commonest of the smaller species is the leopard cat (Colour Plate 2) which has a widespread east and south Asian range. This cat weighs 3–5 kg. It can swim well, and is found on many small coastal islands in the region. It hunts on the ground, and occurs in forest, in settled countryside and in town areas, and can be a nuisance to poultry keepers. It purrs harshly, and utters a loud, raucous mew. I once reared a baby leopard cat with the litter of a house cat. After it had been mixed with her kittens, puss accepted it–albeit uneasily–but she leapt from the box whenever the poor thing mewed or purred. It survived, none the less, but gradually became wild, at first lurking under furniture and ultimately taking to the garden and vanishing.

Other cats have divergent habits. The scarce and little known marbled cat is arboreal, a denizen of forest. Its long tail may help it balance when climbing. The jungle cat, golden cat and its close relative in Borneo, the bay cat, are ground-living–and relatively short-tailed; the first does not penetrate south of the Isthmus of Kra into the ever-wet rain forest zone. The flat-headed cat (Colour Plate 2) and fishing cat are fish-eaters. In both the claws are incompletely retractible.

The gestation periods of Felidae, where known, vary

with body size: 100–108 days for tigers, 90–100 days for leopards, 90–95 days for clouded leopards, about 95 days for the golden cat, and 55–72 days for the leopard cat. One to four is the normal range of litter sizes, up to seven young among tigers. The cubs or kittens are born furred, but helpless, with eyes closed.

6
The Ungulates

WHEN adult, the largest of the hoofed mammals, or ungulates, are generally avoided by predatory carnivores. Their chief enemy is man. The depradations of man have a long history, reaching back centuries to the earliest use of animal parts such as rhinoceros horn or elephant ivory. More significantly, as land clearance and development have advanced, large herbivores have come into conflict with farmers and planters, and have been subjected to control measures while also experiencing wholesale loss of natural habitat. The normal home ranges of those such as elephant, the rhinoceroses, tapir or wild cattle are large, and individuals or herds often hold obstinately to their traditional grounds, despite the removal of original vegetation and other disturbance.

Elephants

Indian elephants are also involved in the special relationship of domestication. In South-East Asia elephants were kept at many royal courts, including Brunei where caparisoned processional elephants were seen by Pigafetta in 1521. In the nineteenth and early twentieth centuries they were much used as working or riding animals: in lumbering, for transport in rough, roadless country and even on the railways. New animals have always been wild-caught; because of the long gestation period (about twenty-one months) and slow growth, captive breeding is not economic. Latterly, at least, the custom of rounding up elephants has continued only in Burma and Thailand. Even when the demand was great, however, the abstraction of animals for this purpose did not seriously deplete the wild population.

Elephants are gregarious, associating in herds comprising a dominant bull and several cows, calves and immature animals of both sexes. Each herd has its own beat, covering many square kilometres and traversed by traditional routes. They often resort to mineral springs which are apparently vital to them as sources of essential elements otherwise deficient in their diets. It has been the conversion of natural habitats to agriculture or plantation, with crops so palatable to elephants, that has brought them into conflict with man, and led to serious depletion or extermination over so much of their South-East Asian range.

Fossil elephants of late Pleistocene age have been found in Java and in north-western Borneo, where none occurred wild in historic times. The suggestion has been made that those now living in eastern Sabah and northern East Kalimantan are re-introductions, descended from domestic elephants released in the mid-eighteenth century. Although the absence of elephants from the rest of Borneo is not satisfactorily explained, it is reasonable to accept that the present population of this area has in fact derived naturally from prehistoric origins.

Censuses of wild elephants have wide margins of error, and may all too quickly be out of date. Estimates published in 1985 are 500–2,000 in Borneo, 796–955 in Peninsular Malaysia and 3,000–6,000 in Burma. No figure later than 1977 is available for Thailand, where the stock was then put at 2,000 head, mostly in protected areas, including Khao Yai National Park, or in border hill country.

Tapirs

The tapirs are ancient and primitive large ungulates, with four toes on the fore-feet and three on the hind, rather long, flexible snouts and short tails. Their close ancestors already existed some thirty million years ago. Fossils indistinguishable from the modern genus *Tapirus*, to which all living forms belong, are known from North America, Europe and continental Asia well into the Pleistocene era (less than two million years ago). Since then, tapirs have vanished over enormous tracts of the world and now survive only in the tropical forests of South America and South-East Asia. Our one species, the Malayan tapir, is therefore of great interest as a relic of this once widespread group.

As an adult boldly patterned black and white (Plate 13) and, like all tapirs, brown with buffish spots and stripes as a young calf, the Malayan tapir feeds by browsing on a mixture of plants, mostly associated with forest gaps, disturbed or secondary growth. Tapirs tend to follow jungle trails. Their tracks and piles of dung (not unlike rhino droppings) show that

13. Tapirs wading in the upper River Sat, Taman Negara, Malaysia.

they travel long distances in all kinds of country, highland and lowland. In primary forest in Malaysia, movements of a radio-tagged male during five months covered 12.7 sq km, overlapping the ranges of others.

Archaeology, once again, has shown that the tapir was present in late Pleistocene Borneo and survived there to at least 6,000 years ago, if not until the recent past. There is no evidence that its extermination was due to prehistoric man; natural causes must have been effective. It is now in decline everywhere, but hunting and habitat loss are undoubtedly the main factors.

Rhinoceroses

Archaeology has also shown that the Javan rhinoceros has disappeared from Borneo since about 12,000 years ago, evidently from natural causes. In Peninsular Malaysia, this species has proved most susceptible to modern pressures, having last been recorded in the 1930s. The present predicament of both rhinoceroses in South-East Asia is chiefly attributable to the high value attached to their horns, in particular, other bodily organs and even blood.

The trade in rhinoceros horn and other parts is ancient. The effects were already notable by the nineteenth century and became acute in the twentieth. In Borneo, a scientific expedition in 1893–4 to the upper Kapuas, West Kalimantan, found hunters active and no signs of rhino and, in 1937, after a prolonged visit to remote and once populous rhino country in the upper Trusan, Sarawak, E. Banks wrote: 'I saw only once a trace made about three years ago . . . every one of the many old wallows passed had the remains of a hut within a few hundred yards, and even on the highest peaks the wandering hunters had left their traces.'

It now seems shocking that in the early 1930s, when it was realized that the Javan rhinoceros might soon be extinct, an expedition was mounted to shoot the last animals for museum specimens. Yet this happened in Perak in 1932, supported by the Game Department. Current debate centres on the merits of live-capture for zoos (and, with luck, captive breeding), as opposed to conservation in the wild. Meanwhile, one protected population of Javan rhinoceroses exists at Ujung Kulon, Java. Groups of Sumatran rhinoceroses survive in reserves including Gunung Leuser, Sumatra, and Taman Negara and Endau–Rompin, Malaysia. The total world population is thought to be less than 300. The problems of conservation are compounded by the large areas of habitat required: densities in primary forest tracts at Endau–Rompin vary from one Sumatran rhinoceros per 40 sq km to one per 120 sq km.

Of the two, the Javan is larger. Its skin bears a surface pattern of mosaic-like roughened thickenings and is folded into stiff, permanent creases. Three such folds cross the back, one before and one behind the shoulders and one over the rump. In the Sumatran rhinoceros, the skin folds are reduced and only two cross the spine, one over the fore-parts and one over the rump. The surface texture of the skin is granular, and it bears a sparse but even coat of short, stiff hairs. Some specimens, especially those in zoos, grow a patchy coat of shaggy hair, which may be tinged reddish brown; others are rubbed bare. On pieces of skin of a freshly killed wild Sumatran rhinoceros from Sabah, which I examined for forensic purposes, the hairs were black and blunt-ended, uniformly abraded to an even length.

Both rhinoceroses are browsers, feeding on a mix of foliage and fruit, particularly of plants associated with disturbed forest or secondary vegetation, such as *Macaranga* spp., *Mallotus* spp., *Artocarpus* spp. and figs. They are normally solitary and sedentary, but not territorial in so far as the home ranges of individuals may overlap. They make wallows in muddy ground, frequent mineral springs and deposit their dung at fixed points where large piles of their nodular droppings accumulate.

Pigs

The pigs are the most successful ungulate group in the forested environments of South-East Asia. Representatives of the family are found throughout the region, including many islands; some island populations have been carried by man beyond their natural range. For many rural communities, wild pigs are the main source of mammalian meat in the diet. A study for World Wildlife Fund Malaysia in 1984–5, led by J. O. Caldecott, concluded that in Sarawak (state area 123,000 sq km, human population 1.3 million) about one million bearded pigs were killed and consumed each year, with 23,000 sambar and 31,000 barking deer. The combined annual value of this meat was M$210 million at prevailing prices up-river, M$320 million if replaced with domestic pork and beef. Similar figures would certainly be forthcoming if such calculations were repeated elsewhere. If properly managed, wild pigs and other game animals can be harvested on a sustainable basis and will then make a major recurrent contribution to rural economies.

All pigs have large heads with long, mobile snouts and powerful jaws equipped with upwardly curving canines and low-crowned molars. They are omnivorous ground-feeders, finding much of their food (animal and vegetable) by rooting in soft soil. Pigs have been called 'the gardeners of the forest'. No doubt, their digging turns and aerates the topsoil but, above a certain density, pigs can prevent the growth of tree seedlings and so interfere with the cycle of natural regeneration.

The most unusual pig is the long-limbed, sparsely haired babirusa, in which the upper canines are completely reversed and grow upwards from their bony sockets through the skin on each side of the snout, curving backwards towards the eyes (Figure 3). The babirusa occurs on Celebes and nearby islands, including Buru in the Moluccas where it may have been introduced. It has no obvious relative among other pigs of the region, all of which are of the one genus, *Sus*.

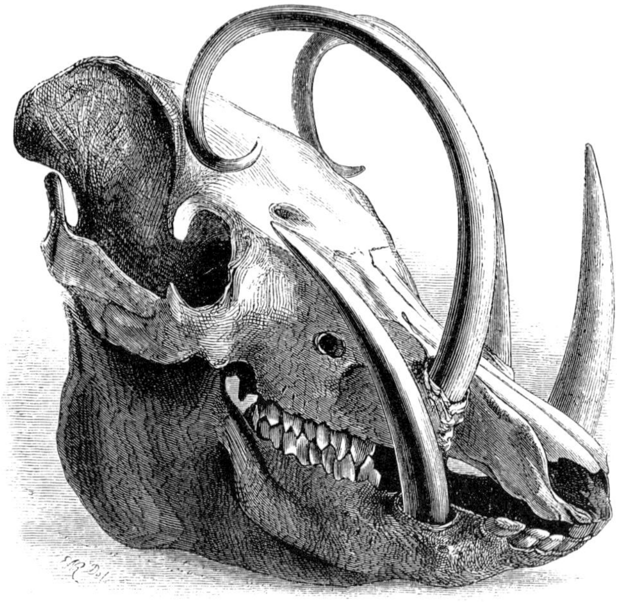

Figure 3. Skull of babirusa.

Among the *Sus* species, the young are striped yellow or buff on brown. This pattern is soon lost, and older pigs are more or less uniformly coloured, black or dark brown, except for the mature bearded pig which is white, with white or rusty red hairs. All build characteristic 'nests' or shelters, large piles of vegetation gathered in the immediate vicinity by the pig, bitten off or uprooted. Stems of grass, tree seedlings, twigs, shrubs and even thorny rattans are used, and the effort in terms of pig-hours is astonishing.

The Eurasian wild pig has an enormous range from the Atlantic to the Pacific. It occurs naturally in South-East Asia through Peninsular Malaysia to Sumatra and Java. This pig, however, is the ultimate ancestor of all domestic varieties, and in this form has been introduced to and gone wild on many other islands. Some introductions were made early in the history of man when the domestic pig was little different from its wild ancestor; others occurred much later from European stock carried by seafarers. Thus, there is dispute whether the pigs of Enggano Island, Sumatra, were brought by man in ancient times or swam ashore from a nineteenth-century shipwreck.

The processes of selection act rapidly in isolation, and the pigs of many South-East Asian islands have proved difficult to assign definitely to one or other of the wild species. For instance, while the pig on Palawan Island is recognized as a variant bearded pig, those of other Philippine islands are strikingly different (Plate 14 (a) and (b)) and are therefore listed in the Checklist (Appendix) under their own name. There is also controversy over the identification of the feral pigs of islands of eastern Indonesia. It has even been suggested that at some past time, the Celebes pig was also domesticated, that the feral pigs of Timur and Roti represent this species rather than *Sus scrofa* and that those of the Moluccas (and the whole island of New Guinea) are descended from a cross between domestic *Sus scrofa* and Celebes pigs. This interesting hypothesis has yet to be tested genetically or immunologically. The Eurasian wild pig occurs naturally alongside the bearded pig in Peninsular Malaysia and Sumatra, and the warty pig in Java. In Java, isolated instances of possible interbreeding have been noted, but over their natural shared ranges, each of these species has remained distinct.

Pigs are gregarious, forming groups that vary with circumstances, the age of the animals, etc. For instance, adult females with small young normally live apart from others. The gestation period of the Eurasian wild pig in Peninsular Malaysia is about 110 days; one to eight (average 4.5) young are born. At about four months they are weaned, whereupon the sow can again become pregnant. Other species show equal fecundity. This high reproductive capacity means that, in favourable circumstances, pig populations can increase rapidly. Apparent-

14. (a) A male and (b) female Philippines pig from the Western Cordillera, Panay, involved in a captive breeding programme.

ly in response to variations in food supplies, bearded pigs travel long distances through the forest. Gregarious movements by bearded pigs in large herds have been reported in Peninsular Malaysia and, especially, in Borneo where huge numbers are sometimes involved. During 1983, for instance, an estimated quarter of a million pigs travelled through the interior Baram district, Sarawak. Many were killed by hunters.

Mouse-deer

The mouse-deer or chevrotains chew the cud but do not have the fully developed, four-chambered stomach characteristic of true ruminants. Their predecessors are known as fossils in North America, Europe and Asia, from some thirty-five million years ago (Oligocene–Miocene). They are now found only in a relic tropical distribution, one species in equatorial Africa, one in southern India and Ceylon (Sri Lanka) and two in South-East Asia. They lack antlers and the deeply recessed scent gland in front of the eye ('lachrymal gland'), found in true deer. The males grow long, sharp, curved upper canine teeth and develop a curious bony plate below the skin over the hind quarters, the function of which is not known.

The mouse-deer are nocturnal, generally solitary except when with young, inhabiting the undergrowth of forest. Fallen fruit are important in their diet. Where both occur, they are readily distinguished by size, the smaller weighing 1–2 kg, the larger 4–6 kg. Populations of intermediate size and atypical coloration occur on islands surrounding Peninsular Malaysia. Evidence including chromosome studies and immunology relates these to the smaller mouse-deer. The gestation period of both species is about five months and both normally give birth to one young. This is weaned at four to five months, and the dam immediately comes into heat again.

Deer

All true deer of South-East Asia have a deep lachrymal gland below the eye, but only barking deer males grow long canines. The males of all carry antlers which are bony outgrowths of the skull, periodically shed and regrown during the stag's lifetime. In the barking deer, the haired extension of the skull from which the antlers grow (the 'pedicel') is long; among the *Cervus* deer it is short. While growing, antlers are covered with closely haired skin (known as 'velvet') supplied with blood and nerves and, in effect, a sensitive part of the living animal. The operation of cropping the half-grown antlers in velvet, quite commonly performed on captive deer, should therefore be done under veterinary supervision with effective anaesthesia.

The barking deer are true forest deer, living solitarily, paired or in mother–young couples. They stand about a half metre high at the shoulder, are generally ruddy brown in colour, showing a white flash on the underside of the raised tail. Their deep barking call, for which they are named, is distinctive. The taxonomy of the genus *Muntiacus* is unsettled. Based on careful scrutiny of museum collections, what was considered to be one widespread species, *M. muntjak*, has been separated into two in Borneo, and the status of *M. feai* is problematical. No ecological differences between these named forms have been found.

The *Cervus* deer are larger, ranging from about 700 cm at the shoulder in the hog deer to 1 400–1 600 cm among sambar. The first coat of the calves is invariably spotted, and some species may also be marked with spots as adults, either permanently in both sexes (spotted deer and Prince Alfred's spotted rusa), or seasonally or in females only. Spotted deer exist in South-East Asia only as introductions, and Javan rusa have also been artificially spread by man beyond their natural range, through the Nusatenggara and Moluccas island groups as far as New Guinea. At higher latitudes sexual activity among these deer is confined to a seasonal rut, and the antlers are shed and regrown in a related cycle. In the equatorial populations of sambar, however, no periodicity has been found in the production of young or the renewal of antlers.

Deer feed by browsing and by grazing. Their favoured habitat is the forest edge, clearings or open swamp or grassy country. Many are gregarious, normally associating in herds. They are thus very susceptible to hunting and habitat loss. All are under pressure and some are severely threatened. Schomburgk's deer originally inhabited the central plains of Thailand, from which they were driven by the expansion of rice cultivation in the second half of the nineteenth century. The species is now almost certainly extinct, remembered only by the trophies of richly branching antlers, some museum specimens and murky, late nineteenth century photographs of zoo animals. Brow-antlered deer are adapted to the open deciduous forests of the central plains of Burma and the undulating hills of interior Cambodia and eastern Thailand. In Burma, specially

15. Prince Albert's spotted rusa, in Panay Island, the Philippines.

created reserves at Shwesettaw and Kyatthin support reasonable populations, but throughout its range there are severe general pressures on this deer. Hog deer have become scarce, and on the Visayan Islands, Philippines, the last few Prince Albert's spotted rusa are subject to an emergency rescue operation (Plate 15 and Frontispiece).

Wild Cattle

South-East Asia is home to seven of the world's eleven species of wild cattle. The horns of cattle, which are present in both sexes, consist of a bony core surrounded by a horny sheath, and are not shed. Bulls are invariably larger than cows. All are grazers and therefore, even more than the deer, dependent on clearings, glades or grassy river banks. In former times, gaur and banteng could be found widely, if patchily, throughout the open country of monsoonal South-East Asia. Kouprey, only discovered in the 1930s, coexisted with the others (and with wild buffalo) in the rolling hills of northern Cambodia and adjoining parts of Thailand, Laos and Vietnam where, unfortunately, war and insurgency have been rife for so long. The present status of kouprey and other wild cattle in this area is unknown.

In the equatorial, mainly island zone, wild cattle find suitable habitat in the lower valleys of large river systems. Gaur (in Peninsular Malaysia and Sumatra) and banteng (in Java and Borneo) overlap in distribution in Burma, Thailand and Cambodia. Both are gregarious, forming herds of up to twenty-five individuals, in which there is usually only one mature bull. 'Bachelor' bulls may be solitary or gather in small herds. They seek shelter in forest by day, emerging to graze at nightfall. The gestation period is about nine months and normally only one calf is born. Gaur are the largest of cattle, old bulls standing up to 195 cm at the prominent shoulder ridge and weighing up to 900 kg. They are interfertile with domestic cattle, and cross-breeding sometimes occurs. Banteng have been domesticated as Bali cattle, and huge numbers now run wild in northern Australia.

The wild water buffalo was also formerly found in the Burma–Thai–Indo-china region, and some may still exist, although their natural habitat has largely been converted to agriculture and they are liable to cross with domestic stock. The tamarau is a half-size buffalo, unique to Mindoro, rare and threatened. The two anoas show characters intermediate between true cattle and buffaloes, but are lightly built, almost deer-like in proportions. The mountain anoa, confined to high elevations in Celebes, stands about 70 cm at the shoulder, has a black or dark brown woolly coat and smooth horns, round in cross-section. The lowland anoa, found in swampy country in the north of Celebes, stands about 85 cm, has a thin coat, black with white stockings, and horns flat in cross-section.

Goat-antelopes

Last of the ungulates are the two goat-antelopes. Both typically favour steep, hilly or rocky terrain. The goral occurs from the Himalayas through the high ground of northern Burma into Thailand. The serow is more widespread in South-East Asia, reaching Peninsular Malaysia and Sumatra.

The serow stands about 90 cm at the shoulder, with a coarse black or grey coat and a mane on the neck. The goral is smaller, about 70 cm at the shoulder, brown in general colour, paler and yellower on the underparts. Both are rare and shy, and under pressure from habitat loss and hunting. The serow is believed to have magical powers to heal its own wounds, simply by licking. By inference, medicinal and therapeutic value is attributed to its flesh and to grease rendered from it. For this reason, the animal is highly valued and eagerly sought by hunters.

7
Anteaters, Rodents and Hares

Anteaters

ALSO under pressure because of imputed medicinal properties are the strangest of South-East Asia's mammals, the pangolins or scaly anteaters (Plate 16). The upper body surface of these mammals is clad in overlapping horny scales which, in their growth and composition, are modified hairs; the underside is sparsely coated in normal hairs. Medicinal powers are attributed to the scales. The raw material of commerce is the dried skin with scales attached. Statistics from a small sector, that is, annual exports through Sarawak, hint at the extent of the trade in Borneo. Quantities peaked in 1962 at 16 tons, approximately equivalent to 11,000 adult pangolins in the year.

The pangolin is an ant- and termite-eater: the stomach of one shot in the wild contained over 200,000 ants and pupae. As adaptations to this diet, its snout is elongated and tapered and its mouth devoid of teeth but provided with a long, protusible tongue which serves to lick up its prey. To break into the nests of these insects, the feet are equipped with strong, curved claws, so large in proportion that the sole of the foot cannot be placed flat on the ground: in effect, a pangolin walks on the sides of its feet. The long tail is prehensile and, with its help, pangolins are skilled climbers.

The Chinese pangolin is found in the northern fringe of South-East Asia. Elsewhere, in the more humid, heavily forested regions of the Sunda Shelf (including Palawan, Culion and Busuanga in the Philippines), the Malaysian pangolin occurs. Because the taxonomy of mammals is so heavily based on dental anatomy, these toothless types are not readily classified and their relationships are uncertain. Early fossils are

16. A female Malaysian pangolin, suckling her young.

not known, but palaeontological finds have shown that a giant pangolin lived in middle Pleistocene Java and survived in Borneo into the last glacial phase, perhaps 30,000 years ago. This was about two and a half times the size of the existing Malaysian pangolin in linear dimensions.

Rodents

The ants and termites of the South-East Asian forests are diverse and numerous. Pangolins are not alone in exploiting this rich resource. The ant-eating model–elongated, tapering snout, long tongue and reduced dentition–is imitated by several unrelated species of ground-dwelling rodents. This example is but one illustration of the amazing variability found among this important mammalian group in the region.

The basic character of rodents, to which they owe their world-wide dominance, lies in the mouth. In each jaw the rodent is provided with a single pair of strong, curved, ever-growing, self-sharpening chisel-like incisors. Behind these, the front premolars at least (and sometimes all) are missing, and there is a wide gap ('diastema') in the toothrow. The insides of the cheeks close across this gap, so that the rodent can continue to use its incisors without losing its mouthful. The molars have a variety of complex folded patterns of the enamel, when in use creating ridged, grinding tooth surfaces. In some groups, such as the voles, the molars elongate continually throughout the animal's life.

With this dental equipment, a rodent can gnaw through hard plant tissue, including the protective coat surrounding many seeds and nuts, and so tap this rich source of nutrition. Some rodents gather more nuts than their immediate needs and store the surplus; they then are acting as dispersal agents to the benefit of the plant. But, ultimately, when the nut is breached, all chances of germination are destroyed and the rodent is thus counted as a seed 'predator'.

Notwithstanding the evident advantages of the rodent dentition in exploiting this otherwise safe and secret food store, several groups have abandoned vegetarianism for an animalivorous way of life. Independent examples are found among both principal South-East Asian families, the squirrels and the rats. All involve convergent modifications towards the ant-eating type. Among the squirrels, progressive examples are found among small ground-dwelling species: the three-striped ground squirrel and the long-nosed mountain ground squirrel have slightly elongated muzzles and mixed diets. The shrew-faced ground squirrel has a longer, narrower snout, short, weak incisors and a long, round, protrusible tongue correlated with a diet in which insects and earthworms predominate. Most extreme is the Celebes long-nosed squirrel, with a very long, narrow snout and weak incisors, strong whiskers and a short tail, correlated with a partially subterranean, burrowing life in the montane forest of Celebes, presumably subsisting off a diet chiefly composed of invertebrate prey. Comparable adaptations are shown by South-East Asian rats, reaching a similar extreme in the Mt. Isarog shrew rat of Luzon, Philippines; but because the rat's skull is naturally elongated, the change in its shape is less striking.

Squirrels

As might be expected in a group of mammals adapted for arboreal life, the South-East Asian squirrel fauna is richest in the region of evergreen tropical rain-forest. There are twenty species of typical squirrel endemic to the Sundaic subregion. In the multi-storeyed forest environment, the squirrels have diversified to fill many niches. Basic adaptations characteristic of all are: the short muzzle and rounded skull, giving good binocular vision in order to judge distances with precision; broad feet with long, strong toes ending in sharp, curved claws by which squirrels can cling to the bark and run up and down, halt, twist and turn on the trunks and branches of trees with facility; and the long, bushy tail which gives balance and also, by erecting or sleeking the hairs and lowering, lifting or flicking with rapid jerks, permits a variety of visual signals. Some squirrels are totally arboreal (for example, the giant squirrels, *Ratufa* spp.), others occupy the middle zone of the forest, the trunks and branches (the pigmy squirrels, *Exilisciurus*) and yet others live mainly on the ground (tufted ground squirrel, shrew-faced ground squirrel). Body weights vary from 15–20 g among pigmy squirrels to 1.5 kg among giant squirrels.

The typical squirrels are active by day. The ground-dwelling forms tend to be cryptic in coloration and behaviour, but in the tree-tops clear signals are important and many arboreal squirrels are brightly coloured and have loud, carrying calls. Geographical barriers, especially the lower courses of large rivers, restrict inter-breeding between arboreal squirrel communities and thus delimit local populations which are distinguished by unique colours or patterns. Extreme examples are the variable squirrel of Thailand, with thirteen recognized subspecies (including five island populations), varying in colour from all black to all white, as well as red-brown and olivaceous, and Prevost's squirrel, with fifteen subspecies in Borneo alone (Colour Plate 4), six of them island populations.

Their poor ability to cross geographical barriers is reflected in the restricted distribution of squirrels on islands beyond the limits of the Sunda Shelf. In the Philippines, typical squirrels are found only on Palawan and the southern islands, represented by endemic species of two Sundaic genera, namely pigmy squirrels *Exilisciurus* and the tree squirrels *Sundasciurus*. On mainland Celebes, apart from the anomalous record of Prevost's squirrel, there is a small fauna of five species in three genera which are endemic, that is to say, exclusive to the island and found nowhere else. In the northern Philippines and east of Celebes, no squirrels occur.

Flying Squirrels

The typical squirrels are matched by an equally diverse fauna of flying squirrels (Colour Plate 5). The two groups are separated only at subfamily level. In fact, differences in the mode of attachment of the patagium and in the anatomy of the skull suggest that there may be two or more distinct lines of descent within the flying squirrels. The flying squirrel fauna is again richest in the humid forested environment, with eleven Sundaic endemics and totals of fourteen species in Borneo and eleven in Peninsular Malaysia. Two genera have independently colonized parts of the Philippines, *Hylopetes* on Palawan and *Petinomys* on Mindanao and adjoining islands. None has reached Celebes or any other eastern island.

As already noted (Chapter 2), gliding flight is an adaptation to nocturnal life, and flying squirrels conform. They too range in size from the minute up to body weights of 1.5 kg. Their diets are poorly known, because observation is difficult, but seem to be similar to those of typical squirrels. Indeed, at a fruiting forest tree, it is often possible at dusk to watch the typical squirrels retire and soon afterwards to see the flying squirrels glide in to resume the feast.

In the rain-forest, squirrels are important primary consumers, individually more numerous than primates. As with primates, the environment is partitioned among the community by subtle differences in the composition of diets and in the vertical stratum of forest most frequented. As noted above (Chapter 4), at Kuala Lompat, Kerau Game Reserve, Malaysia, the average density of gibbons and monkeys (six species) collectively was about 180 per sq km, while typical squirrels (nine species) amounted to 486 individuals per sq km. However, in terms of aggregate weight ('biomass'), the disparity is reversed: primates 835 kg, squirrels 130 kg per sq km. Three species of flying squirrel were known at the study site, but no reliable data were obtained on numbers or biomass.

Squirrels of both kinds build nests among the twigs or in tree-holes. Breeding is seasonal at higher latitudes, where there may also be an annual moult to a brighter pelage. Among equatorial populations, there is no general periodicity. Reproductive rates are not high. In general, litters of one or two young are common and four the maximum recorded for any rain-forest squirrel.

Bamboo Rats

Père David's vole is the only representative within South-East Asian bounds of a family widespread in the temperate regions of the world. In this case, it inhabits mountain forests. The bamboo rats, by contrast, are a unique South-East Asian group, extending outside our region no further than southern China and related only to the root rats of Africa. They show general adaptations to a burrowing life: a stout, fusiform body, broad skull for the attachment of strong jaw musculature, short ears, short, powerful limbs and short, naked tail. The feet are equipped with strong claws. Bamboo rats live largely in underground burrows from which they feed on subterranean roots and storage organs, notably the rhizomes of bamboos. They occur only in forests or forest edge habitat where these food sources exist, but can be pests of crops such as manioc or sugar cane. They come out above ground by night and, if surprised, freeze in an intimidating open-mouthed posture, displaying their truly formidable, orange-edged incisors and coughing hoarsely.

Rats and Mice

The rats and mice of the family Muridae are the most numerous, varied and successful small mammals of South-East Asia, a world centre of their diversity. The genus of typical rats, *Rattus*, has now spread to all parts of the globe but probably originated here. The area is also especially rich in endemic genera and species. More than any other rodent group, or other small mammals such as shrews or treeshrews, the murids have crossed the sea barrier to colonize the many islands of the region. With a short generation time, they have proved highly responsive to the selective influences of isolation and the absence of predators and potential competitors.

Some have been assisted by man, commensal with him, dependent on the artificial environments created by his activities and often gravely injurious to his aspirations. The bandicoot rats, house mouse, house rat and Norway (brown) rat are town and household pests; the rice-field rats and mice, Polynesian rat and Malaysian wood rat infest fields and plantations, damaging crops and stored products. These familiar rats and mice are universally detested and controlled where possible by trap, poison or any other means to hand. Yet it would be tragic if experience of a handful of pests was to taint our appreciation of the 150 or so other species in over forty genera, constituting the true, native wild murid fauna of South-East Asia.

Rats and mice are variously coloured, mostly shades of light or dark brown or grey-brown, often paler or white on the underparts (Colour Plate 6). The pelage may be soft and woolly, or coarse and rough, or spiny, or a mixture of hairs and spines. Smallest among South-East Asian species is probably the grey-bellied pencil-tailed tree-mouse of Borneo, head and body 66–80 mm long, plus tail 85–97 mm; largest probably the giant rat of Flores, head and body 410–450 mm, tail 330–370 mm.

The basic shape of the rat is adaptable to many modes of existence. The snout is comparatively long and the body fusiform. The gait is four-footed, but the fore-feet have good grasping powers with the four clawed toes (plus a vestigial first) while the hind feet are broad and long, with five toes, and a hairless sole bearing small, rounded pads. The tail is long and covered in small scales which are arranged in rings and interspersed with minute hairs or bristles.

Rats living on the forest floor have moderate to long scaly tails, and hind feet in which the sole is comparatively long and narrow and the toes short and slender in proportion (for example, the spiny rats, *Maxomys*). Many rats have exploited the three-dimensional nature of the forest environment and become scansorial. Typical climbing adaptations parallel those already identified among squirrels (above) and include fore-shortening of the muzzle, to give improved binocular vision, and modifying the feet, extending and strengthening the toes while reducing the overall length of the foot. Some climbing rats have strong claws on partially opposable inner and outer toes of the foot. Among others, the claws are small, and grip is apparently obtained by means of thickened pads on the sole of the foot. In the marmoset rat, which has minute claws, permanently raised and out of contact with the substrate, the sweat glands in the skin of the lower surface of the foot are prominent and active, secreting a copious watery fluid. The rat lives among bamboos, on which it is dependent both for food and for nest sites. It runs freely up and down the smooth outer surface of the largest bamboo stems, leaving small damp toe-prints as it goes.

Climbing rats generally have long tails, which are used to give counter-pressures and support, but are not prehensile (Plate 17). There is also a tendency to hairiness on the tail. The usual

17. A marmoset rat *Hapalomys longicaudatus* feeds on the terminal shoots of the large bamboo, *Gigantochloa scortechinii*; note the grasping feet and scaly, bristly tail, used as a support but not prehensile.

form this takes is a weak bottle-brush at the tip, as in the marmoset rat and pencil-tailed tree-mouse (Plate 18). In the Luzon cloud rat (*Phloeomys*) the entire tail is covered in short hairs. The ultimate is reached by the bushy-tailed rats (*Crateromys*) (Figure 4). The Luzon species, *C. schadenbergi*, with its immensely thick, bushy tail, displays several adaptations strikingly convergent with squirrels (which are, of course, absent from the northern Philippines).

There is much to be discovered about the biology of these fascinating mammals. It is assumed that all rats have a gestation period of three weeks, but this has rarely been tested among

18. A pencil-tailed tree mouse *Chiropodomys gliroides* at the entrance to its nest in a bamboo stem.

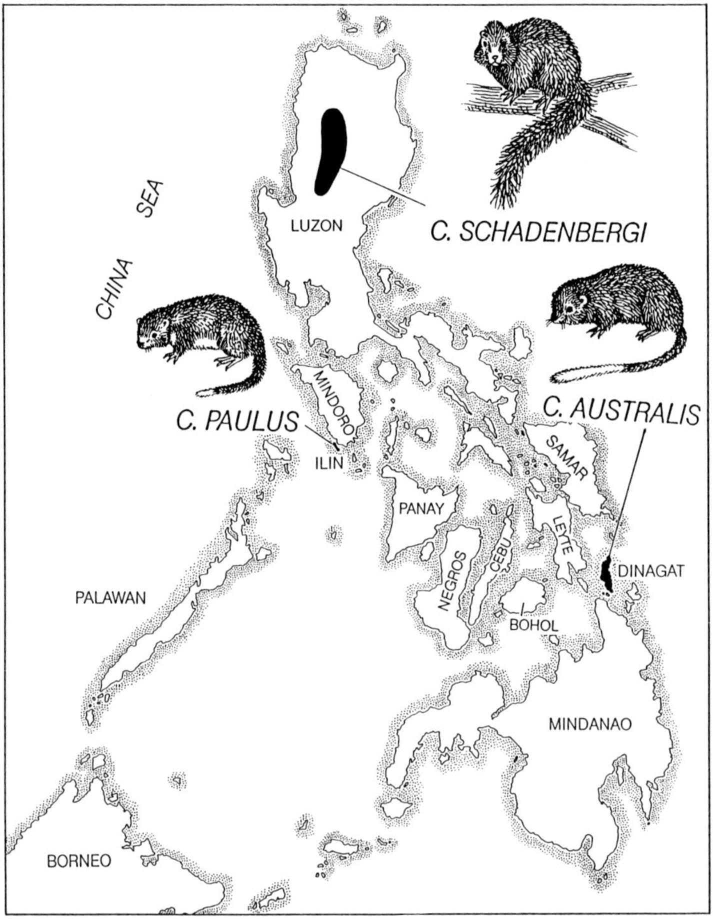

Figure 4. The distribution of the three species of bushy-tailed rats.

non-pest species. The litter sizes of many are known only from the uterine contents of trapped females. For the pencil-tailed tree-mouse of Peninsular Malaysia the figure is one to four young, average 2.4 (20 instances). For larger, ground-dwelling forest rats the range is greater (one to seven young) and averages higher: Whitehead's spiny rat *Maxomys whiteheadi* 3.0, long-tailed giant rat *Leopoldamys sabanus* 3.1, red spiny rat *Maxomys surifer* 3.3. Among the pest species, *Rattus* spp., local records of litter sizes range up to ten or eleven, and the averages are higher at 5.5. Data such as these are non-existent for many species, especially the island rats. Too many of them are known only from the dried skins and skulls of small series of specimens collected decades ago. The vulnerability of natural habitat on small South-East Asian islands is so acute that many populations must be threatened, if not already lost.

Porcupines

The porcupines, for the time being, appear to be safe from such anxieties. They are the largest of South-East Asian rodents. The short-tailed and thick-spined porcupines, with long black and white spines, are familiar (Plate 19). The brush-tailed porcupine

19. A short-tailed porcupine.

20. A long-tailed porcupine.

has a duller coloration and smaller spines. The trend is continued in the local endemic long-tailed porcupine (Plate 20) which is almost rat-like in general appearance.

The dentition consists of typical chisel-shaped rodent incisors and cheek teeth with an intricately folded enamel, giving a convoluted pattern of whorls and circles on the worn chewing surface. Porcupines are mixed vegetarians, feeding on fallen fruit, shoots, bark and tree saps, but not above chewing a bone if it is available.

Hares

Finally, to complete this survey, brief mention must be made of the Indian hare, introduced and apparently naturalized in Java, and the Burmese hare, inhabiting the open country of the monsoon region of South-East Asia. These two are representatives of a group very widespread in the grasslands of Europe and Asia. Quite different is the Sumatran hare. This is the only species of its genus, confined to the mountains of Sumatra, forest-dwelling, the only hare or rabbit with a striped coat, believed to live in burrows but to be incapable of digging its own, solitary, nocturnal. What a mysterious creature! How much there remains to be discovered about the lives of this and so many other South-East Asian mammals.

Checklist

THIS nominal list of the mammals of South-East Asia is the basis of the book and sets the order of progression of the text. The scientific nomenclature largely follows the compilation of Honacki, Kinman and Koeppl (1982), amended to take account of subsequent publications noted in the bibliography. The English vernacular names are those used in the text, guided chiefly by van Strien (1986), Boonsong and McNeely (1977), Morris (1965) and Walker (1964), but occasionally by invention.

GEOGRAPHICAL RANGES can only be indicated in a generalized manner. Numbers are used to denote occurrence within regional, national or island bounds, as shown on the map (endpapers). For the purpose of this list, 'South-East Asia' is defined as the national territories of Burma, Laos, Vietnam, Thailand, Cambodia, Malaysia, Singapore, Brunei Darussalam, the Philippines and Indonesia excluding the province of Irian Jaya. Thus Bangladesh, India and China (including Hainan and other islands) are extralimital (1), as are the Tanimbar, Kei, Aru and Waigeo island groups, all New Guinea and Australia (14).

As far as can be ascertained, the list includes all species of mammals occurring in this region, omitting only a few of primarily extralimital distribution on continental Asia whose ranges extend marginally into northern parts of South-East Asia, as here defined.

Distribution within the specified territories of South-East Asia is indicated by numerals, as follows: (2) Burma; (3) Thailand; (4) the Indo-chinese states, i.e., Laos, Vietnam and Cambodia; (5) Peninsular Malaysia, or (5+) if also extending northwards beyond the national boundary towards a limit at the southern tip of peninsular Burma and the neighbourhood of Chumphon in peninsular Thailand, i.e. encompassing the region traditionally known as the 'Malay peninsula'. (6) Sumatra; Nias, Batu and/or one or more of the Mentawai Islands separately identified as (6a), and remote Enggano (6b). Borneo and any of is associated islands, (7). Java (8) may include Madura; tenggara, from Lombok to Savu and Alor are (9), and Timur (10). (11) denotes Celebes (Sulawesi) with Selayar, Butung and Sula Island; (12) the Philippines, with certain restricted ranges indicated as (12a) the Palawan or (12b) the Mindanao faunal areas, or named island(s). The Moluccas (13) may be subdivided into North (13a) and South (13b). Throughout certain instances of localized occurrence, on a single island or island group, are given in parentheses – ( ) – after the encompassing regional number, while occurrences attributed to artificial introduction, or for some other reason of uncertain status, are enclosed in square brackets – [ ].

It must be emphasized that this listing cannot show whether a mammal is widespread or localized, common or rare in any particular part of its range. A broad brush has to be applied. Readers to whom details of range are important will need to refer to national or regional sources, such as U Tun Yin (1967) for Burma, Boonsong Lekagul and McNeely (1977) for Thailand, Medway (1983) for Peninsular Malaysia, Payne, Francis and Phillipps (1985) for Borneo, van Strien (1986) for Indonesia and Heaney (1986) for the Philippines, all of which have been consulted in the preparation of the list.

The sequence of families follows systematic order. Genera within families and species within genera are listed alphabetically by Latin name.

MARSUPIALIA

Peramelidae		
Seram bandicoot	*Rhynchomeles prattorum*	13 (Seram)
Phalangeridae		
Small cuscus	*Phalanger celebensis*	11
Spotted cuscus	*P. maculatus*	13b, 14
Grey cuscus	*P. orientalis*	10, 13b, 14
Bear cuscus	*P. ursinus*	11
Petauridae		
Sugar glider	*Petaurus breviceps*	13a, 14

INSECTIVORA

Erinaceidae		
Gymnure or moonrat	*Echinosorex gymnurus*	5, 6, 7
Lesser gymnure	*Hylomys suillus*	1, 2, 3, 4, 5, 6, 7. 8
Dinagat gymnure	*Podogymnura aureospinula*	12 (Dinagat)
Mindanao gymnure	*P. truei*	12b
Shrew-hedgehog	*Neotetracus sinensis*	1, 2, 3, 4
Soricidae		
Chinese short-tailed shrew	*Anourosorex squamipes*	1, 2, 3
Red-toothed shrew	*Soriculus salenskii*	1, 2, 3
Sunda water shrew	*Chimarrogale phaeura*	5, 6, 7

White-toothed shrews	*Crocidura attentuata*	1, 2, 3, 4, 5
	C. horsfieldi	1, 3, 4
	C. fuliginosa	1, 2, 3, 4, 5, 6, 7, 10, 11
	C. monticola	3, 5, 7, 8, 9
	C. maxi	8
	C. russula	1, 3, 4
Celebes shrews	*C. elongata*	11
	C. lea	11
	C. levicula	11
	C. nigripes	11
	C. rhoditis	11
Philippine shrews	*C. beatus, C. grandis*	12b
	C. parvacauda	12b
	C. grayi	12 (Luzon)
	C. halconus, C. mindorus	12 (Mindoro)
	C. edwardsiana	12 (Sulu)
	C. negrina	12 (Negros)
	C. palawanensis	12 (Palawan)
Kinabalu shrew	*Suncus ater*	7
Pigmy musk shrew	*Suncus etruscus*	1, 2, 3, 5, 7
Flores shrew	*Suncus mertensis*	9 (Flores)
House shrew	*Suncus murinus*	Widespread
Philippine musk shrews	*Suncus luzonensis*	12
	Suncus occultidens	12
	Suncus palawanensis	12a

Talpidae

Short-tailed mole	*Talpa micrura*	1, 2, 3, 4, 5

SCANDENTIA

Tupaiidae

Northern smooth-tailed treeshrew	*Dendrogale murina*	3, 4
Bornean smooth-tailed treeshrew	*Dendrogale melanura*	7
Pentail	*Ptilocercus lowii*	5+, 6, 7
Striped treeshrew	*Tupaia dorsalis*	7
Common treeshrew	*Tupaia glis*	1, 2, 3, 4, 5, 6, 7, 8
Slender treeshrew	*Tupaia gracilis*	7
Indonesian treeshrew	*Tupaia javanica*	6, 8, 8b
Lesser treeshrew	*Tupaia minor*	5+, 6, 7
Mountain treeshrew	*Tupaia montana*	7
Palawan treeshrew	*Tupaia palawanensis*	12a
Painted treeshrew	*Tupaia picta*	7
Ruddy treeshrew	*Tupaia splendidula*	7
Large treeshrew	*Tupaia tana*	6, 7
Mindanao treeshrew	*Urogale everetti*	12b

CHECKLIST

DERMOPTERA

Cynocephalidae		
Flying lemurs or	*Cynocephalus variegatus*	2, 3, 4, 5, 6, 7, 8
Colugo	*Cynocephalus volans*	12

CHIROPTERA
MEGACHIROPTERA

Pteropodidae		
(Pteropodinae)		
Celebes flying foxes	*Acerodon arquatus*	11
	A. celebensis	11
Talaud flying fox	*A. humilis*	Talaud Is., Indonesia
Philippines flying foxes	*A. jubatus*	12
	A. leucotis	12a
	A. lucifer	12
Sunda flying fox	*A. mackloti*	9, 10
Grey fruit bat	*Aethalops alecto*	5, 6, 7, 8
Katanglad fruit bat	*Alionycteris paucidentata*	12b
Spotted-winged fruit bat	*Balionycteris maculata*	5+, 7
Manado flying fox	*Boneia bidens*	11
Black-capped fruit bat	*Chironax melanocephalus*	5+, 6, 7, 8, 11
Dog-faced fruit bats	*Cynopterus archipelagus*	12 (Polillo)
	C. brachyotis	1, 2, 3, 4, 5, 6, 7, 11
	C. horsfieldi	5+, 6, 7, 8
	C. sphinx	1, 2, 3, 4, 5, 6, 8
	C. terminus	10
	C. titthaecheileus	6, 8, 8a, 9 (Lombok)
Bare-backed fruit bats	*Dobsonia exoleta*	11
	D. moluccensis	13b, 14
	D. peroni	9, 10
	D. viridis	11, 13
Dayak fruit bat	*Dyacopterus spadiceus*	5, 6, 7
Mindoro fruit bat	*Haplonycteris fischeri*	12 (Mindoro)
Tail-less fruit bats	*Megaerops ecaudatus*	1, 3, 4, 5, 6, 7
	M. kusnotoi	8
	M. wetmorei	12
Small-toothed flying fox	*Neopteryx frosti*	11
Luzon fruit bat	*Otopteropus cartalagonodus*	12 (Luzon)
Dusky fruit bat	*Penthetor lucasi*	5, 7
Philippine fruit bats	*Ptenochirus jagori*	12
	P. minor	12
Central flying fox	*Pteropus alecto*	9, 11, 13
Silver flying fox	*P. argentatus*	13b (Ambon)
	P. balutus	12 (Sarangani Is.)
Ashy-headed flying fox	*P. caniceps*	13
Ambon flying fox	*P. chrysoproctus*	13b

Spectacled flying fox	*P. conspicillatus*	13a (Halmahera)
Grey flying fox	*P. griseus*	9, 10, 11
Island flying fox	*P. hypomelanus*	3, 4, 5, 6, 7, 8, 9, 10, 11, 13, 14
	P. leucopterus	12
Lombok flying fox	*P. lombocensis*	9
Lyle's flying fox	*P. lylei*	3, 4
Mearns' flying fox	*P. mearnsi*	12
Black-bearded flying fox	*P. melanopogon*	13b
Black-eared flying fox	*P. melanotus*	1, 6a, 6b
Seram flying fox	*P. ocularis*	13b
Masked flying fox	*P. personatus*	11, 13a
Miangas flying fox	*P. pumilus*	Taulad, Miangas I.
Sebutu flying fox	*P. speciosus*	12 (Sebutu)
	P. tablasi	12
Temminck's flying fox	*P. temminckii*	13b
Large flying fox	*P. vampyrus*	5+, 6, 7, 8, 9, 12
Rousettes	*Rousettus amplexicaudatus*	2, 3, 5, 6, 7, 8, 9, 11, 12, 13, 14
	R. celebensis	11
	R. leschenaultii	1, 2, 3, 4, 8, 8a
	R. spinalatus	6, 7
Blanford's bat	*Sphaerias blanfordi*	1, 2, 3
Stripe-faced flying fox	*Styloctenium wallacei*	11
Swift fruit bat	*Thoopterus nigrescens*	11, 13a
(Macroglossinae)		
Dulit fruit bat	*Eonycteris major*	6a, 7, 12
Cave fruit bat	*E. spelaea*	1, 2, 3, 4, 5, 6, 7, 8, 9, 10, 11
Long-tongued fruit bats	*Macroglossus minimus*	3, 4, 5, 6, 7, 8, 8a, 11, 12, 14
	M. sobrinus	1, 2, 3, 4, 5, 6, 8
Blossom bats	*Syconycteris australis*	13b, 14
	S. carolinae	13a (Halmahera)
(Nyctimeninae)		
Tube-nosed fruit bats	*Nyctimene aello*	13, 14
	N. albiventer	13a, 14
	N. cephalotes	10, 11, 13
	N. minutus	11, 13b
(Harpionyterininae)		
Harpy fruit bats	*Harpionycteris celebensis*	11
	H. whiteheadi	12

MICROCHIROPTERA

Rhinopomatidae		
Mouse-tailed bat	*Rhinopoma microphyllum*	1, 3, 6

Emballonuridae		
Sheath-tailed bats	*Emballonura alecto*	7, 11, 12
	E. monticola	3, 5, 6, 7, 8, 11
	E. nigrescens	11, 13, 14
	E. raffrayana	13b, 14
White-bellied tomb bat	*Saccolaimus saccolaimus*	1, 3, 5, 6, 7 8, 9, 14
Long-winged tomb bat	*Taphozous longimanus*	1, 2, 3, 4, 5, 6, 7, 8, 9
Black-bearded tomb bat	*T. melanopogon* (inc. *philippinensis*)	1, 2, 3, 4, 5, 6, 7, 8, 9, 10, 11, 12
Theobald's tomb bat	*T. theobaldi*	1, 2, 3, 4, 8
Craseonycteridae		
Hog-nosed bat	*Craseonycteris thonglongyai*	3
Nycteridae		
Hollow-faced bats	*Nycteris javanica*	8, 8a
	N. tragata	5+, 7
Megadermatidae		
Large false vampire	*Megaderma lyra*	1, 2, 3, 5
Lesser false vampire	*M. spasma*	1, 2, 3, 4, 5, 6, 7, 8, 11, 12, 13
Rhinolophidae		
(Hipposiderinae)		
Trident-nosed bats	*Aselliscus stoliczkanus*	1, 2, 3, 4, 5
	A. tricuspidatus	13b, 14
Tail-less horseshoe bats	*Coelops frithi*	1, 2, 3, 4, 5, 8, 8a
	C. robinsoni (inc. *hirsutus*)	3, 5, 12 (Mindoro)
Roundleaf horseshoe bats	*Hipposideros armiger*	1, 3, 4, 5
	H. ater	1, 5, 7, 12, 13, 14
or Leaf-nosed bats	*H. bicolor*	1, 2, 3, 4, 5, 6, 7, 8, 9, 10, 11
	H. breviceps	6a
	H. cervinus	5, 6, 7, 11, 12, 13, 14
	H. cineraceus	1, 2, 3, 4, 5, 7
	H. coronatus	12b
	H. coxi	7
	H. diadema	2, 3, 4, 5, 6, 7, 8, 11, 12, 13, 14
	H. dinops	11, 14
	H. dyacorum	7
	H. fulvus	1, 2, 3, 4
	H. galeritus	1, 3, 5, 6, 7, 8
	H. inexpectatus	11
	H. larvatus	1, 2, 3, 4, 5, 6, 7, 8, 9
	H. lekaguli	3, 5
Shield-faced bat	*H. lylei*	2, 3, 5

Round-leaf horseshoe bats	*H. nequam*	5
	H. obscurus	12
	H. pygmaeus	12
	H. ridleyi	5, 7
	H. sabanus	5, 6, 7
	H. turpis	1, 3
(Rhinolophinae)		
Horseshoe bats	*Rhinolophus acuminatus*	3, 4, 6, 7, 8, 9, 12
	R. affinis	1, 2, 3, 4, 5, 6, 8, 9
	R. anderseni	12 (Luzon)
	R. arcuatus	7, 11, 12, 13b
	R. borneensis	4, 7, 8, 8a
	R. coelophyllus	2, 3, 5
	R. celebensis	11
	R. creaghi	7, 8, 9
	R. euryotis	13, 14
	R. feae	2, 3
	R. importunus	8
	R. inops	12b
	R. keyensis	13, 14
	R. lepidus	1, 2, 3, 5, 6
	R. luctus	1, 2, 3, 4, 5, 6, 7, 8, 8a
	R. macrotis	1, 2, 3, 4, 5, 6, 12
	R. madurensis	8 (Madura)
	R. malayanus	3, 4, 5
	R. marshalli	3
	R. nereis	Anamba & N. Natuna Is.
	R. paradoxolophus	3, 4
	R. pearsoni	1, 2, 3, 4
	R. philippinensis	7, 10, 11, 12
	R. pusillus	1, 2, 3, 4, 5, 6, 7, 8
	R. robinsoni	3, 5
	R. rufus	12
	R. sedulus	5, 7
	R. shameli	2, 3, 4, 5
	R. simplex	9
	R. stheno	3, 5, 6, 8
	R. subrufus	12
	R. thomasi	2, 3, 4
	R. toxopeusi	13
	R. virgo	12
	R. yunanensis	1, 3
Vespertilionidae		
(Vespertilioninae)		
Serotines	*Eptesicus demissus*	1, 3
	E. pachyotis	1, 2, 3
	E. serotinus	1, 3

Thick-thumbed pipistrelles	*Glischropus javanus*	8
	G. tylopus	2, 3, 5, 6, 7, 12, 13
False serotines	*Hesperoptenus blanfordi*	3, 5+, 7
	H. doriae	5, 7
	H. gaskelli	11
	H. tickelli	1, 2, 3
	H. tomesi	5, 7
Great evening bat	*Ia io*	1, 3, 4
Mouse-eared bats	*Myotis adversus*	1, 5, 6, 7, 8, 11, 13, 14
	M. annectans	1, 2, 3
	M. ater	7, 11, 13, 14
	M. chinensis	1, 3
	M. formosus	1, 6, 8, 8a, 11, 12
	M. hasseltii	1, 2, 3, 4, 5, 6, 7, 8
	M. horsfieldii	1, 3, 5, 6, 7, 8, 8a, 11
	M. macrotarsus	7, 12
	M. montivagus	1, 2, 5, 7
	M. muricola	1, 2, 4, 5, 6, 7, 8, 11, 12
	M. ridleyi	5, 6, 7
	M. rosseti	3, 4
	M. siligorensis	1, 3, 4, 5
	M. stalkeri	13
Thick-thumbed bat	*Philetor brachypterus*	5, 6, 7, 8, 12, 14
Pipistrelles	*Pipistrellus cadornae*	1, 2, 3
	P. ceylonicus	1, 2, 4, 7
	P. circumdatus	1, 2, 5, 8
	P. cuprosus	7
	P. imbricatus	7, 8, 8a, 12
	P. javanicus	1, 2, 3, 4, 5, 6, 7, 8, 9, 11, 12
	P. kitcheneri	7
	P. lophurus	2
	P. macrotis	5, 6
	P. mimus	1, 2, 3, 4
	P. minahassae	11
	P. mordax	1, 8
	P. peguensis	2
	P. petersi	7, 11, 13
	P. pulveratus	1, 3
	P. societatis	5
	P. stenopterus	5, 6, 7, 12
	P. tenuis	5+, 6, 7, 8, 11, 12, 13, 14
	P. vordermanni	7, Belitung
Harlequin bat	*Scotomanes ornatus*	1, 2, 3, 4
Yellow bats or House bats	*Scotophilus celebensis*	11
	S. heathi	1, 2, 3, 4
	S. kuhli	1, 2, 3, 4, 5, 6, 7, 8, 9, 12, 13, 14

Flat-headed bats	*Tylonycteris pachypus*	1, 2, 3, 4, 5, 6, 7, 8, 9
	T. robustula	1, 2, 3, 4, 5, 6, 7, 8, 11
(Miniopterinae)		
Bent-winged bats	*Miniopterus australis*	7, 8, 11, 12, 13, 14
	M. magnater	1, 2, 3, 5, 6, 7, 8
	M. medius	3, 5, 8, 14
	M. pusillus	1, 3, 5, 6, 8, 11, 13, 14
	M. schreibersii	1, 2, 3, 4, 5, 6, 7, 8, 9, 11, 13, 14
	M. tristis	11, 12, 14
(Murininae)		
Hairy-winged bats	*Harpiocephalus harpia*	1, 2, 4, 7, 8, 13b
	H. mordax	2, 7
Tube-nosed bats	*Murina aenea*	5, 7
	M. aurata	1, 2, 3
	M. balstoni	8
	M. canescens	6a
	M. cyclotis	1, 2, 3, 4, 5, 7
	M. florium	9, 11, 13b, 14
	M. huttoni	1, 2, 3, 4, 5
	M. rozendaali	7
	M. suilla	5, 6, 7, 8
	M. tubinaris	1, 2, 3, 4
(Kerivoulinae)		
Funnel-eared bats	*Kerivoula hardwickei*	1, 2, 3, 4, 5, 6, 7, 8, 9, 11, 12
	K. intermedia	7
	K. minuta	5+, 7
	K. papillosa	1, 4, 5, 6, 7, 8, 11
	K. pellucida	5, 6, 7, 8, 12
Painted bat	*K. picta*	1, 2, 3, 4, 5, 6, 7, 8, 9, 13
	K. whiteheadi	5+, 7, 12
Groove-toothed bats	*Phoniscus atrox*	5+, 6, 7
	P. jagori	7, 8, 8a, 11, 12
Molossidae		
Meldrum's bat	*Chaerephon johorensis*	5, 6
Wrinkle-lipped bat	*C. plicata*	1, 2, 3, 4, 5, 6, 7, 8, 12
Hairless bat	*Cheiromeles torquatus*	5+, 6, 7, 8, 11, 12
Free-tailed bats	*Mops mops*	5, 6, 7
	M. sarasinorum	11, 12
	Mormopterus beccarii	13b, 14
	M. doriae	6
	Otomops formosus	8

CHECKLIST

PRIMATES

Lorisidae		
Slow lorises	*Nycticebus coucang*	1, 2, 3, 4, 5, 6, 7, 8, 12b
	N. pygmaeus	4
Tarsiidae		
Tarsiers	*Tarsius bancanus*	6, 7
	T. spectrum	11
	T. syrichta	12
Cercopithecidae		
(Cercopithecinae)		
Stump-tailed macaque	*Macaca arctoides*	1, 2, 3, 4
Assamese macaque	*M. assamensis*	1, 2, 3, 4
Long-tailed macaque	*M. fascicularis*	2, 3, 4, 5, 6, 7, 8, 12
Celebes [Moor] macaque	*M. maura*	11
Rhesus macaque	*M. mulatta*	1, 2, 3, 4
Pig-tailed macaque	*M. nemestrina*	1, 2, 3, 4, 5, 6, 7
Celebes macaque	*M. nigra*	11
Celebes [Booted] macaque	*M. ochreata*	11
Celebes [Tonkean] macaque	*M. tonkeana*	11
(Colobinae)		
Proboscis monkey	*Nasalis larvatus*	7
Javan grey langur	*Presbytis comata*	8
Silvered langur	*P. cristata* (= *aurata*)	2, 3, 4, 5, 6, 7, 8, 8a, 9 (Lombok)
Francois' langur	*P. francoisi*	1, 4
White-fronted langur	*P. frontata*	7
Hose's langur	*P. hosei*	7
Banded langur	*P. melalophos*	2, 3, 5, 6, 7
Dusky langur	*P. obscura*	2, 3, 5
Phayre's langur	*P. phayrei*	1, 2, 3, 4
Mentawai langur	*P. potenziani*	6a
Maroon langur	*P. rubicunda*	7
Thomas's langur	*P. thomasi*	6
Douc langur	*Pygathrix nemaeus*	4
Snub-nosed monkey	*Rhinopithecus avunculus*	4
Pig-tailed langur	*Simias concolor*	6a
Hylobatidae		
Agile gibbon	*Hylobates agilis*	5+, 6, 7
Concolor gibbon	*H. concolor*	1, 4
Hoolock gibbon	*H. hoolock*	1, 2
Kloss gibbon	*H. klossi*	6a
White-handed gibbon	*H. lar*	2, 3, 5, 6
Grey gibbon	*H. moloch*	8
Borneo gibbon	*H. muelleri*	7
Pileated gibbon	*H. pileatus*	3, 4
Siamang	*H. syndactylus*	5, 6

Pongidae		
Orang-utan	*Pongo pygmaeus*	6, 7

CARNIVORA

Canidae		
Jackal	*Canis aureus*	1, 2, 3
Dhole	*Cuon alpinus*	1, 2, 3, 4, 5, 6, 8
Ursidae		
Sun bear	*Helarctos malayanus*	1, 2, 3, 4, 5, 6, 7
Black bear	*Selenarctos thibetanus*	1, 2, 3, 4
Mustelidae		
(Mustelinae)		
Yellow-throated marten	*Martes flavigula*	1, 2, 3, 4, 5, 6, 7, 8
Java weasel	*Mustela lutreolina*	6, 8
Malay weasel	*M. nudipes*	5+, 6, 7
Siberian weasel	*M. sibirica*	1, 2, 3, 4
Striped weasel	*M. strigidorsa*	1, 2, 3
(Melinae)		
Hog badger	*Arctonyx collaris*	1, 2, 3, ?5, 6
Ferret badgers	*Melogale moschata*	1, 2, 3, 4
	M. personata	1, 2, 3, 4, 7, 8
Teledu	*Mydaus javanensis*	6, 7, 8
Philippine badger	*M. marchei*	12a
(Lutrinae)		
Small-clawed otter	*Amblonyx cinerea*	1, 2, 3, 4, 5, 6, 7, 8, 12
Eurasian otter	*Lutra lutra*	1, 2, 3, 4, 6, ?7, 8
Smooth otter	*L. perspicillata*	1, 2, 3, 4, 5, 6, 7
Hairy-nosed otter	*L. sumatrana*	4, 5+, 6, 7, 8
Viverridae		
Binturong	*Arctictis binturong*	1, 2, 3, 4, 5, 6, 7, 8, 12a
Three-striped palm civet	*Arctogalidia trivirgata*	1, 2, 3, 4, 5, 6, 7, 8
Banded civet	*Chrotogale owstoni*	1, 4
Otter civet	*Cynogale bennettii*	4, 5+, 6, 7
Banded civet	*Hemigalus derbyanus*	2, 3, 5, 6, 7
Hose's civet	*H. hosei*	7
Brown palm civet	*Macrogalidia musschenbroekii*	11
Masked palm civet	*Paguma larvata*	1, 2, 3, 4, 5, 6, 7, [12a]
Common palm civet	*Paradoxurus hermaphroditus*	1, 2, 3, 4, 5, 6, 7, 8, 12 [9, 10, 11, 13 ?intr.]
Banded linsang	*Prionodon linsang*	2, 3, 5, 6, 7, 8
Spotted linsang	*P. pardicolor*	1, 2, 3, 4
Large-spotted civet	*Viverra megaspila*	1, 2, 3, 4, 5

Malay civet	*V. tangalunga*	5, 6, 7, 11, 12, 13b
Indian civet	*V. zibetha*	1, 2, 3, 4, 5
Small Indian civet	*Viverricula indica*	1, 2, 3, 4, 5, 6, 8, 8a [9 (Sumbawa) ?intr.]
Herpestidae		
Short-tailed mongoose	*Herpestes brachyurus*	5, 6, 7, 12a
Hose's mongoose	*H. hosei*	7
Javan mongoose	*H. javanicus*	1, 2, 3, 4, 5, 8
Collared mongoose	*H. semitorquatus*	6, 7
Crab-eating mongoose	*H. urva*	1, 2, 3, 4 [12a]
Felidae		
Bay cat	*Felis badia*	7
Leopard cat	*F. bengalensis*	1, 2, 3, 4, 5, 6, 7, 8, 12a
Jungle cat	*F. chaus*	1, 2, 3, 4
Marbled cat	*F. marmorata*	1, 2, 3, 4, 5, 6, 7
Flat-headed cat	*F. planiceps*	5+, 6, 7
Golden cat	*F. temminckii*	1, 2, 3, 4, 5, 6
Fishing cat	*F. viverrina*	1, 2, 3, 4, 6, 8
Clouded leopard	*Neofelis nebulosa*	1, 2, 3, 4, 5, 6, 7
Leopard	*Panthera pardus*	1, 2, 3, 4, 5, 8
Tiger	*P. tigris*	1, 2, 3, 4, 5, 6, 8 (8a extinct)

PROBOSCIDEA

Elephantidae		
Indian elephant	*Elephas maximus*	1, 2, 3, 4, 5, 6, 7

PERISSODACTYLA

Tapiridae		
Malayan tapir	*Tapirus indicus*	2, 3, 4, 5, 6
Rhinocerotidae		
Sumatran rhinoceros	*Dicerorhinus sumatrensis*	2, 3, 4, 5, 6, 7
Javan rhinoceros	*Rhinoceros sondaicus*	1, 2, 3, 4, 5, 6, 8

ARTIODACTYLA

Suidae		
Babirusa	*Babyrousa babyrussa*	11, 13 (Buru)
Bearded pig	*Sus barbatus*	5, 6, 7, 12a
Celebes pig	*S. celebensis*	11 [10 intr.]
Philippines pig	*S. philippinensis*	12
Eurasian wild pig	*S. scrofa*	1, 2, 3, 4, 5, 6, 8
Warty pig	*S. verrucosus*	8

Tragulidae		
Lesser mouse-deer	*Tragulus javanicus*	2, 3, 4, 5, 6, 7, 8
Large mouse-deer	*T. napu*	2, 3, 4, 5, 6, 7
Cervidae		
Prince Alfred's spotted rusa	*Cervus alfredi*	12
[Spotted deer]	*C. axis*	[8 intr.]
Calamianes deer	*C. calamianensis*	12a
Brow-antlered deer	*C. eldi*	2, 3, 4
Mariannas deer	*C. marianus*	12
[Sika]	*C. nippon*	[12 (Sulu) intr.]
Hog-deer	*C. porcinus*	1, 2, 3, 4
Schomburgk's deer	*C. schomburgki*	(3 extinct)
Javan rusa	*C. timorensis*	8, 9, 10, 11 [7 intr.]
Sambar	*C. unicolor*	1, 2, 3, 4, 5, 6, 7
Barking deer	*Muntiacus atherodes*	7
	M. feai	2, 3
	M. muntjak	1, 2, 3, 4, 5, 6, 7, 8, 8a [9 (Lombok) intr.?]
Bovidae		
Lowland anoa	*Anoa depressicornis*	11
Mountain anoa	*A. quarlesi*	11
Gaur	*Bos frontalis*	1, 2, 3, 4, 5
Banteng	*B. javanicus*	2, 3, 4, 7, 8
Kouprey	*B. sauveli*	3, 4
Water buffalo (wild)	*Bubalus bubalis*	1, 3, 4
Tamarau	*B. mindorensis*	12 (Mindoro)
Serow	*Capricornis sumatraensis*	1, 2, 4, 5, 6
Goral	*Nemorhaedus goral*	1, 2, 3

PHOLIDOTA

Manidae		
Malaysian pangolin	*Manis javanica*	2, 3, 4, 5, 6, 7, 8, 12
Chinese pangolin	*M. pentadactyla*	2, 3, 4

RODENTIA

Sciuridae		
(Sciurinae)		
Ear-spot squirrel	*Callosciurus adamsi*	7
Kloss squirrel	*C. albescens*	6
Kinabalu squirrel	*C. baluensis*	7
Grey-bellied squirrel	*C. caniceps*	2, 3, 5
Pallas's squirrel	*C. erythraeus*	1, 2, 4
Variable squirrel	*C. finlaysoni*	?2, 3, 4
Belly-banded squirrel	*C. flavimanus*	1, 2, 3, 4, 5
Mentawai squirrel	*C. melanogaster*	6a
Black-banded squirrel	*C. nigrovittatus*	4, 5+, 6, 8
Plantain squirrel	*C. notatus*	5+, 6, 7, 8 [11 (Selayar) intr.?]

Borneo black-banded squirrel	*C. orestes*	7
Prevost's squirrel	*C. prevostii*	5+, 6, 7, ?11
Irrawaddy squirrel	*C. pygerythrus*	1, 2, 4
Long-nosed mountain squirrels	*Dremomys everetti*	7
	D. pernyi	4
	D. rufigenis	2, 3, 4, 5
Pigmy squirrels	*Exilisciurus concinnus*	12
	E. exilis	7
	E. luncefordi	12b
	E. samaricus	12 (Samar, Leyte)
	E. whiteheadi	7
Sculptor squirrels	*Glyphotes canalvus*	7
	G. simus	7
Celebes long-nosed squirrel	*Hyosciurus heinrichi*	11
Four-striped ground squirrel	*Lariscus hosei*	7
Three-striped ground squirrel	*L. insignis*	5+, 6, 7, 8
Indochinese ground squirrel	*Menetes berdmorei*	2, 3, 4
Black-eared pigmy squirrel	*Nannosciurus melanotis*	6, 7, 8
Celebes dwarf squirrels	*Prosciurillus abstrusus*	11
	P. leucomus	11
	P. murinus	11
Common giant squirrel	*Ratufa affinis*	2, 3, 4, 5, 6, 7
Black giant squirrel	*R. bicolor*	1, 2, 3, 4, 5, 6, 8, 8a
Tufted ground squirrel	*Rheithrosciurus macrotis*	7
Shrew-faced ground squirrel	*Rhinosciurus laticaudatus*	5+, 6, 7
Celebes red-bellied squirrel	*Rubrisciurus rubriventer*	11
Brooke's squirrel	*Sundasciurus brookei*	7
Horse-tailed squirrel	*S. hippurus*	5+, 6, 7
Busuanga squirrel	*S. hoogstraali*	12 (Busuanga)
Jentink's squirrel	*S. jentinki*	7
Low's squirrel	*S. lowii*	5+, 6, 7
Philippine squirrels	*S. philippinensis* (inc. *davensis, mindanensis, samarensis*)	12
	S. steerii (inc. *albicauda, juvencus, mollendorffi*)	12
Palawan squirrel	*S. rabori*	12a
Slender squirrel	*S. tenuis*	5+, 6, 7
Striped tree squirrels	*Tamiops macclellandi*	1, 2, 3, 5
	T. maritimus	1, 4
	T. rodolphei	3, 4
	T. swinhoei	2, 4
(Petauristinae)		
Black flying squirrel	*Aeromys tephromelas*	5+, 6, 7
Thomas's flying squirrel	*A. thomasi*	7
Hairy-footed flying squirrel	*Belomys pearsoni*	1, 2, 3, 4
Particolored flying squirrel	*Hylopetes alboniger*	1, 2, 3, 4
Red-cheeked flying squirrel	*H. lepidus*	3, 4, 5, 6, 7, 8
Palawan flying squirrel	*H. nigripes*	12a
Phayre's flying squirrel	*H. phayrei*	2, 3

Sipora flying squirrel	*H. sipora*	6a
Grey-cheeked flying squirrel	*H. spadiceus*	5+, 6, 7, 8
Horsfield's flying squirrel	*Iomys horsfieldii*	5, 6, 7, 8
Pigmy flying squirrels	*Petaurillus emiliae*	7
	P. hosei	7
	P. kinlochii	5
Red & white giant flying squirrel	*Petaurista alborufus*	1, 2, 3
Spotted flying squirrel	*P. elegans*	1, 2, 3, 4, 5, 6, 7, 8
Red giant flying squirrel	*P. petaurista*	1, 2, 3, 4, 5, 6, 7, 8
Bartels' flying squirrel	*Petinomys bartelsi*	8
Basilan flying squirrel	*P. crinitus*	12
Whiskered flying squirrel	*P. genibarbis*	5, 6, 7, 8
Hagen's flying squirrel	*P. hageni*	6, 7
White-bellied flying squirrel	*P. setosus*	2, 3, 5, 6, 7
Vordermann's flying squirrel	*P. vordermanni*	2, 5, 6
Smoky flying squirrel	*Pteromyscus pulverulentus*	5, 6, 7
Arvicolidae		
Père David's vole	*Eothenomys melanogaster*	1, 2, 3, 4
Rhizomyidae		
Bay bamboo rat	*Cannomys badius*	1, 2, 3, 4
Hoary bamboo rat	*Rhizomys pruinosus*	1, 2, 3, 4, 5
Chinese bamboo rat	*R. sinensis*	1, 2, 4
Large bamboo rat	*R. sumatrensis*	1, 2, 3, 4, 5, 6
Muridae		
Luzon tree rat	*Abditomys latidens*	12 (Luzon)
Mindoro rat	*Anonymomys mindorensis*	12 (Mindoro)
Philippine long-tailed rats	*Apomys abrae*	12 (Luzon)
	A. datae	12 (Luzon)
	A. hylocetes	12b
	A. insignis	12b
	A. littoralis	12
	A. microdon	12
	A. musculus	12
	A. sacobianus	12 (Luzon)
Mt. Isarog shrew rat	*Archboldomys luzonensis*	12 (Luzon)
Bandicoot rats	*Bandicota bengalensis*	1, 2, 6, 8 [5 intr.]
	B. indica	1, 2, 3, 4, 6, 8 [5 intr.]
	B. savilei	2, 3, 4
Luzon forest rats	*Batomys dentatus*	12 (Luzon)
	B. granti	12 (Luzon)
	B. salomonseni	12b
Lesser white-toothed rat	*Berylmys berdmorei*	2, 3, 4
Bowers's rat	*B. bowersi*	1, 2, 4, 5, 6
Kenneth's rat	*B. mackenziei*	1, 2, 3
Philippine mountain rats	*Bullimus bagobus*	12b
	B. luzonicus	12 (Luzon)

Celebes forest rats	*Bunomys andrewsi*	11
	B. chrysocomus	11
	B. fratorum	11
	B. penitus	11
Fruit rats	*Carpomys melanurus*	12 (Luzon)
	C. phaeurus	12 (Luzon)
Shrew-like rat	*Celaenomys silaceus*	12 (Luzon)
Fea's tree rat	*Chiromyscus chiropus*	2, 3, 4
Pencil-tailed tree mice	*Chiropodomys calamianensis*	12a
	C. gliroides	1, 2, 3, 4, 5, 6, 7, 8, 8a
	C. karlkoopmani	6a
	C. major	7
	C. muroides	7
Luzon striped rats	*Chrotomys mindorensis*	12
	C. whiteheadi	12 (Luzon)
Bushy-tailed rats	*Crateromys australis*	12 (Dinagat)
	C. paulus	12 (Ilin)
	C. schadenbergi	12 (Luzon)
Swamp rats	*Crunomys celebensis*	11
	C. fallax	12 (Luzon)
	C. melanius	12b
	C. rabori	12 (Leyte)
Celebes spiny shrew-rat	*Echiothrix leucura*	11
Celebes grey rat	*Eropeplus canus*	11
Pigmy tree rats	*Haeromys margarettae*	7
	H. minahassae	11
	H. pusillus	7
	H. sp.	12a
Marmoset rats	*Hapalomys delacouri*	1, 4
	H. longicaudatus	2, 3, 5
Sody's tree rat	*Kadarsanomys sodyi*	8
Rinca rat	*Komodomys rintjanus*	10 (Rinca & Padar)
Trefoil-toothed giant rat	*Lenomys meyeri*	11
Grey tree rat	*Lenothrix canus*	5, 6a, 7
Edwards' giant rat	*Leopoldamys edwardsi*	1, 2, 3, 4, 5, 6
Neill's rat	*L. neilli*	3
Long-tailed giant rat	*L. sabanus*	1, 3, 4, 5, 6, 7, 8
Mentawai long-tailed giant rat	*L. siporanus*	6a
Mindanao mountain rat	*Limnomys sibuanus*	12b
Celebes tree rats	*Margaretamys beccarii*	11
	M. elegans	11
	M. parvus	11
Spiny rats	*Maxomys alticola*	7
	M. baeodon	7
	M. bartelsii	8
	M. dollmani	11
	M. hellwaldi	11
	M. hylomyoides	6

	M. inas	5
	M. inflatus	6
	M. moi	4
	M. musschenbroeki	11
	M. ochraceiventer	7
	M. pagensis	6a
	M. panglima	12a
	M. rajah	5+, 6, 7
	M. surifer	2, 3, 4, 5, 6, 7, 8
	M. whiteheadi	5+, 6, 7
Celebes vole rat	*Melasmothrix naso*	11
Banana rats	*Melomys aerosus*	13 (Seram)
	M. fraterculus	13 (Seram)
	M. fulgens	13 (Seram, Talaud Is.)
	M. obiensis	13 (Obi)
Mice	*Mus caroli*	1, 3, 4, 5, 6, 7, 8, 9 (Flores)
Asian house mouse	*M. castaneus*	Towns of the region
	M. cervicolor	1, 2, 3, 4, 6, 8
	M. cookii	1, 2, 3, 4
	M. crociduroides	6
	M. pahari	1, 4
	M. shortridgei	2, 3, 4
	M. vulcani	8
White-bellied rats	*Niviventer bukit*	2, 3, 5, 6, 8, 8a
	N. cremoriventer	5+, 6, 7, 8, 8a
	N. fulvescens	1, 4
	N. hinpoon	3
	N. langbianis	2, 3, 4
	N. lepturus	8
	N. rapit	5, 6, 7
	N. tenaster	2, 3
Palawan rat	*Palawanomys furvus*	12a
Flores giant rat	*Papagomys armandvillei*	10 (Flores)
Celebes rat	*Paruromys dominator*	11
Luzon cloud rats	*Phloeomys cumingi*	12 (Luzon, Mariduque)
	P. pallidus	12 (Luzon)
Red tree rats	*Pithecheir melanurus*	6, 8
	P. parvus	5
Typical rats	*Rattus adustus*	6b
	R. annandalei	5, 6
Rice-field rat	*R. argentiventer*	3, 4, 5, 6, 7, 8, 9, 11, 12, 14
	R. atchinus	6
Summit rat	*R. baluensis*	7
	R. blangorum	6
	R. bontanus	11

	R. callitrichus	11
	R. ceramicus	13 (Seram)
	R. dammermani	11
	R. elphinus	11 (Talubu)
	R. enganus	6b
	R. everetti	12
Polynesian rat	*R. exulans*	2–13, 14
	R. felicius	13 (Seram)
	R. foramineus	11
	R. hamatus	11
	R. hoffmanni	11 (Togian)
	R. hoogerwerfi	6
	R. hoxaensis	4
	R. korinchi	6
	R. losea	1, 3, 4
	R. lugens	6a
	R. marmosurus	11
	R. mindorensis	12 (Mindoro)
	R. morotaiensis	13 (Gilolo gp)
	R. nitidus	1, 2, 3, 12
Norway rat	*R. norvegicus*	Ports
	R. osgoodi	4
	R. punicans	11
House rat	*R. rattus*	Throughout
	R. remotus	3 (Phangan, Tao, Samui)
	R. salocco	11
	R. sikkimensis	1, 2, 3, 4
	R. simalurensis	6a
	R. taerae	11
	R. tawitawiensis	12 (Tawitawi)
Malaysian wood rat	*R. tiomanicus*	5+, 6, 7, 8, 12a
	R. tyrannus	12 (Ticao)
	R. xanthurus	11
Philippines shrew rats	*Rhynchomys isarogensis*	12 (Luzon)
	R. soricoides	12 (Luzon)
Mountain giant rat	*Sundamys infraluteus*	6, 7
Bartels' rat	*S. maxi*	8
Mueller's giant rat	*S. muelleri*	2, 3, 5, 6, 7, 12a
Celebes woolly rats	*Taeromys arcuatus*	11
	T. celebensis	11
Long-footed rat	*Tarsomys apoensis*	12b
Celebes shrew rat	*Tateomys rhinigradoides*	11
Mearn's Luzon rat	*Tryphomys adustus*	12 (Luzon)
Long-tailed cane mouse	*Vandeleuria oleracea*	1, 2, 4
Hystricidae		
Brush-tailed porcupine	*Atherurus macrourus*	1, 2, 3, 4, 5, 6
Short-tailed porcupine	*Hystrix brachyura*	1, 2, 3, 4, 5, 6, 7
Sunda porcupine	*H. javanica*	8, 8a, 9 (Flores), ?11

Thick-spined porcupines	*Thecurus crassispinis*	7
	T. pumilis	12a
	T. sumatrae	6
Long-tailed porcupine	*Trichys fasciculata*	5, 6, 7

LAGOMORPHA

Leporidae

Indian hare	*Lepus nigricollis*	1 [8 intr.]
Burmese hare	*L. peguensis*	2, 3, 4
Sumatran hare	*Nesolagus netscheri*	6

Select Bibliography

Boonsong Lekagul, and McNeely, J. A., *Mammals of Thailand*, Bangkok, Association for the Conservation of Wildlife, 1977.

Cao Van Sung, 'Inventaire des rongeurs du Vietnam', *Mammalia*, Vol. 48, pp. 391–5, 1984.

Davison, G. W. H., 'New records of Peninsular Malaysian and Thai shrews', *Malay. Nat. J.*, Vol. 36, pp. 211–15, 1984.

Groves, C. P., and Grubb, P., 'The species of muntjac (genus *Muntiacus*) in Borneo: unrecognised sympatry in tropical deer', *Zool. Meded., Leiden*, Vol. 56, No. 17, pp. 203–16, 1982.

Grubb, P., and Groves, C. P., 'Notes on the taxonomy of the deer (Mammalia, Cervidae) of the Philippines', *Zool. Anz. Jena*, Vol. 210, Nos. 1/2, Suppl., pp. 119–44, 1983.

Heaney, L. R., 'Systematics of Oriental pigmy squirrels of the genera *Exilisciurus* and *Nannosciurus* (Mammalia: Sciuridae)', *Misc. Papers Mus. Zool., Univ. Michegan*, No. 170, pp. 1–58, 1985.

______, 'Biogeography of mammals in SE Asia: estimates of rates of colonization, extinction and speciation', *Biol. J. Linn. Soc., Lond.*, Vol. 28, pp. 127–65, 1986.

Hill, J. E., 'Bats (Mammalia: Chiroptera) from Indo-Australia', *Bull. Br. Mus. Nat. Hist. (Zool.)*, Vol. 45, No. 3, pp. 103–208, 1983.

______, and Songsakdi Yenbutra, 'A new species of the *Hipposideros bicolor* group (Chiroptera: Hipposideridae) from Thailand', *Bull. Br. Mus. Nat. Hist. (Zool.)*, Vol. 47, pp. 77–82, 1984.

______, and Francis, C. M., 'New bats (Mammalia: Chiroptera) and new records of bats from Borneo and Malaya', *Bull. Br. Mus. Nat. Hist. (Zool.)*, Vol. 47, pp. 305–29, 1984.

______, and Francis, C. M., 'A review of Bornean *Pipistrellus* (Mammalia: Chiroptera)', *Mammalia*, Vol. 50, pp. 43–56, 1986.

______, Zubaid, A., and Davison, G. W. H., '*Hipposideros lekaguli*, a new leaf-nosed bat recorded in Peninsular Malaysia', *Malay. Nat. J.*, Vol. 39, pp. 147–8, 1985.

Honacki, J. H., Kinman, K. E., and Koeppl, J. W., *Mammal species of the world: a taxonomic and geographic reference*, Lawrence, Kansas, USA, Allen Press Inc. & Association of Systematics Collections, 1982.

Jenkins, P. D., 'A discussion of Malayan and Indonesian shrews of the genus *Crocidura* (Insectivora: Soricidae)', *Zool. Meded., Leiden*, Vol. 56, No. 21, pp. 267–79, 1982.

———, and Hill, J. E., 'The status of *Hipposideros galeritus* Cantor, 1846 and *Hipposideros cervinus* (Gould, 1854) (Chiroptera: Hipposideridae)', *Bull. Br. Mus. Nat. Hist. (Zool.)*, Vol. 41, pp. 279–94, 1981.

Medway, Lord, *The Wild Mammals of Malaya (Peninsular Malaysia) and Singapore*, 2nd edn. (reprinted with corrections), Kuala Lumpur, Oxford University Press, 1983.

Morris, D., *The mammals: a guide to the living species*, London, Hodder & Stoughton, 1965.

Musser, G. G., 'A new genus of arboreal rat from Luzon island in the Philippines', *Am. Mus. Novit.*, No. 2730, pp. 1–23, 1982.

———, 'The definition of *Apomys*, a native rat of the Philippine Islands', *Am. Mus. Novit.*, No. 2746, pp. 1–43, 1982.

———, '*Crunomys* and the small-bodied shrew rats native to the Philippine Islands and Sulawesi (Celebes)', *Bull. Am. Mus. Nat. Hist.*, Vol. 174, No. 1, pp. 1–95, 1982.

———, and Califa, Devra, 'Identities of rats from Pulau Maratua and other islands off East Borneo', *Am. Mus. Novit.*, No. 2726, pp. 1–30, 1982.

———, and Heaney, L. R., 'Philippine *Rattus*: a new species from the Sulu archipelago', *Am. Mus. Novit.*, No. 2818, pp. 1–32, 1982.

———, Heaney, L. R. and Rabor, D. S., 'Philippine rats: a new species of *Crateromys* from Dinagat Island', *Am. Mus. Novit.*, No. 2821, pp. 1–25, 1985.

———, and Newcomb, C., 'Definitions of Indochinese *Rattus losea* and a new species from Vietnam', *Am. Mus. Novit.*, No. 2814, pp. 1–32, 1985.

Napier, P. H., and Groves, C. P., '*Simia fascicularis* Raffles, 1821 (Mammalia, Primates): request for the suppression under the plenary powers of *Simia aygula* Linnaeus, 1758, a senior synonym', *Bull. Zool. Nomencl.*, Vol. 40, No. 2, pp. 117–18, 1983.

Payne, J., Francis, C. M., and Phillipps, K., *A field guide to the mammals of Borneo*, Kota Kinabalu, The Sabah Society, 1985.

Strien, N. J. van, *Abbreviated checklist of the mammals of the Australasian archipelago*, Bogor, Indonesia, School of Environmental Conservation Management, 1986.

Tranier, M., 'Un *Glyphotes canalvus* Moore 1959 dans les collections du Muséum National d'Histoire Naturelle (Rodentia, Sciuridae)', *Mammalia*, Vol. 49, pp. 294–6, 1985.

U Tun Yin, *Wild animals of Burma*, Rangoon, Rangoon Gazette, 1967.

Walker, E. P., *Mammals of the world*, Baltimore, USA, Johns Hopkins Press, 1964.

Weitzel, V., and Groves, C. P., 'The nomenclature and taxonomy of the colobine monkeys of Java', *Intl. J. Primatol.*, Vol. 6, pp. 399–409, 1985.

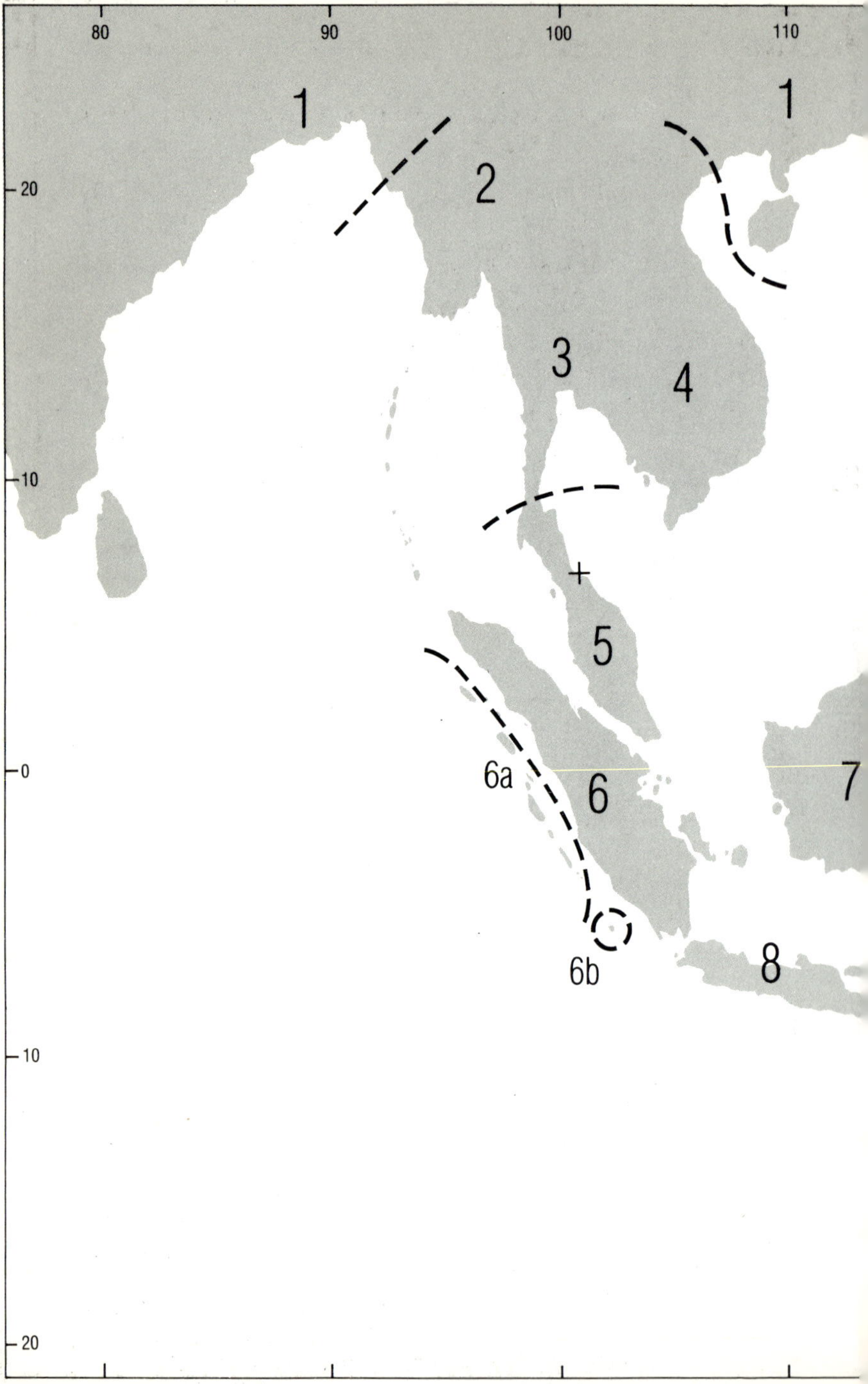

80
90
100
110
20
10
0
10
20
1
1
2
3
4
5
6
6a
6b
7
8